THE NEW POSSIBLE

THE NEW POSSIBLE

Visions of Our World beyond Crisis

Edited by

Philip Clayton, Kelli M. Archie, Jonah Sachs, and Evan Steiner

Foreword by Kim Stanley Robinson

CASCADE *Books* • Eugene, Oregon

THE NEW POSSIBLE
Visions of Our World beyond Crisis

Cascade Books
An Imprint of Wipf and Stock Publishers
199 W. 8th Ave., Suite 3
Eugene, OR 97401

www.wipfandstock.com

PAPERBACK ISBN: 978-1-7252-8583-5
HARDCOVER ISBN: 978-1-7252-8582-8
EBOOK ISBN: 978-1-7252-8584-2

Cataloging-in-Publication data:

Names: Clayton, Philip, 1956–, editor. | Archie, Kelli M., editor. | Sachs, Jonah, editor. | Steiner, Evan, editor. | Robinson, Kim Stanley, foreword.

Title: The new possible : visions of our world beyond crisis / edited by Philip Clayton, Kelli M. Archie, Jonah Sachs, and Evan Steiner.

Description: Eugene, OR: Cascade Books, 2021. | Includes bibliographical references.

Identifiers: ISBN: 978-1-7252-8583-5 (paperback). | ISBN: 978-1-7252-8582-8 (hardcover). | ISBN: 978-1-7252-8584-2 (ebook).

Subjects: LCSH: Human ecology. | Ecotheology. | Climate change. | Ecology. | Social justice.

Classification: GF80 N15 2021 (print). | GF80 (epub).

Manufactured in the U.S.A.   SEPTEMBER 29, 2020

Contents

EDUCATION

LOVE

COMMUNITY

TOMORROW

Illustrations, Figures, and Tables

Foreword

Kim Stanley Robinson

With the appearance of the worldwide COVID-19 pandemic there is now broad general recognition that we have entered an emergency century, a crux in human history. Whether we respond well or poorly will have huge ramifications for future generations of people and for the biosphere at large. From our current moment we could be initiating a mass extinction event that will hammer the biosphere and civilization both, or we could be starting the process of establishing a prosperous and just global society that will be sustainable over the long haul of the centuries to come. The radical disparity of these possible futures, the sheer range of them—but with a kind of excluded middle, in that if we trend in one direction or other that trajectory is likely to prevail—is part of the feeling of our time, which could be characterized as a general sense of danger, dread, and fear, mixed with a battered but still strong feeling of hope that our rapidly increasing scientific knowledge and technological capability, and a rising awareness of our global collective fate, and our ultimate reliance on Earth's biosphere will combine to usher in a new and better era in human interactions with the planet and other people. We live in this curious mixture of fear and hope; probably this has always been the case for humanity, but now it has bloomed into an obvious existential and historical crisis.

The problems we face now are immense and numerous. We are a global society, but we are ruled by a nation-state system in which many still regard national interests as overriding any global considerations. And we have agreed to rule ourselves and run our affairs by way of a political economy that is unsustainable, extractive, and unjust, and yet is massively entrenched in national laws and international treaties. So the nation-state system is insufficient, and yet all we have; and neoliberal capitalism is cruel and destructive, and yet the world's current overriding system of laws. How then to proceed? We have to use the tools at hand, and yet they are precisely a big part of the problem. It's a dilemma.

One thing that may help to start our thinking here is the simple principle that what can't happen, won't happen. This is to invoke the reality principle in the form of the facts of science that are incontrovertible. Magic doesn't work, so magical thinking is not going to be sufficient; physically impossible things are not going to happen in this century or any other, and so we are not going to be conducting our civilization as we have been into the future, because that isn't physically possible. The planet's biosphere doesn't produce the resources we need at the rate we are using them, nor is it capable of disposing of the toxic wastes we are producing at the rate we are producing them. So change will be coming, one way or another, and because the current situation is so very untenable, the changes coming are going to be profound. We are now already in the time of change.

So, in this very perilous situation, we need plans. That's the important task that this book is joining. The editors have divided the general problem and its possible solutions into particular categories and asked experts in these fields to explore their visions of change for the better. As the general problem is a wicked problem, in the technical sense of being multiplex and intractable, the solutions are therefore going to be complex and various. It makes good sense to examine different aspects of the situation in the way this volume does, and it's both helpful and encouraging to see so many voices orchestrated into a larger vision of change that can guide and sustain us through the time of troubles and conflict. The moral obligation to keep hope alive is much aided by some needed clarity about the particulars of how we can deal with the dangers, and transform society for the good. One of the best things about this book is the way the whole adds up to more than the sum of its parts, and becomes that elusive vision of how things could go well, articulated but vibrantly whole.

The group of writers assembled here is very experienced, capable, and articulate, and I recommend all of them to you enthusiastically. In this general excellence I also want to express a particular personal thanks to Vandana Shiva, whom I met almost thirty years ago at a conference organized by Ernest Callenbach and Fritjof Capra, and whose work has since for me set the standard for how a public intellectual can change the world for the better. In my own work I've been trying to complement the efforts of Shiva and all like-minded scientists and educators; I've been trying to tell their stories in utopian fictions, and I couldn't have done it without the inspirational example of people doing the far harder work of changing the real world.

So, what you'll read in this collection of essays is not only inspirational, but interesting; and this is crucial. It's the particulars that are always the interesting part. We are going to have to join in an effort that will be a political battle every step of the way, because unfortunately not everyone is going to see the problem in the same way and not everyone will agree on the solutions, even when the problems become undeniable. So these detailed and meticulous visions for positive change will be very useful going forward, precisely because they are interesting. And they add up to a vision. That's what you've got in this volume, so I give you the joy of it, with many thanks to its editors, publishers, and contributors. Go little book . . .

Kim Stanley Robinson, August 2020

Editors' Introduction

Ours is a time of profound change, perhaps far more than we realize. What we don't yet see, historians will: early 2020 ushered in a season of unprecedented global disruption. In the short-term, we're aware of COVID-19 infection and death rates, protests in the streets, the growing recognition that Black lives matter. But the full ripple effect of these months will only become visible as the economic impacts that are now being felt move completely out into the open. Who can really grasp what it means that *half a billion* people on our planet are expected to fall below the poverty line as a consequence of what has happened this year?

This moment presents each one of us with the question: *How do we come to understand what this all means? Where do we turn? Who can help us to frame appropriate and compassionate responses?*

Even more urgently: *Will humanity fall back into the same unsustainable patterns that made us vulnerable to collapse in the first place? Or can we leverage this crisis as an opportunity to rethink the systems and structures of contemporary society? In what ways can we learn from this profound tragedy how to begin to transition toward a more just and sustainable future?*

Soon after the crisis hit, we asked global leaders with diverse expertise to share what they are learning and feeling during this time of struggle, and what they are now thinking may be possible for the future. The response was staggering: again and again, the top figures in each field agreed to set aside other obligations in order to share their deepest visions with you. Each author describes how they envision the kinds of communities and societies we are now able to build, given the experiences of the last few months. We believe that the "visions of our Earth beyond crisis" that they offer to you here will challenge you to dream in powerful new ways about what is possible . . . and then act to make it so.

The diversity of sources is mind-blowing. Among the participant-observers, Rebecca Kiddle writes of the contribution of Māori lifeways,

Atossa Soltani of the native peoples of the Amazon Basin, Mark Anielski of the economic wisdom of the First Peoples of Canada, Arturo Escobar of the indigenous peoples of Columbia, and Francis Deng of the Dinka culture in South Sudan. The route to the "new possible" clearly leads back through the wisdom and lifeways of this planet's first peoples.

The book rings with contrasts and consonances. Start with the decades of activism that the courageous anti-apartheid activist Mamphela Ramphele brings, and compare her reflections to the thoughts of sixteen-year-old activist Anisa Nanavati, who boldly states her message: "The future belongs to us; here's what we're going to need." Move from the synthetic idea of "ecological civilization" with which Jeremy Lent opens the book, to the specific recommendations on childhood education and gender relations that are among Riane Eisler's "cornerstones for social change," and then on to Tristan Harris's compelling suggestions for re-aligning technology with core human needs.

If "how we've always done it" just isn't going to work anymore, we're going to need some serious new sources of wisdom. It's encouraging, and rare, to find such a broad range of wisdom traditions singing in harmony within a single volume: from the half a dozen different indigenous traditions just mentioned, to the revolutionary manifesto for an "integral ecology" from Pope Francis (Fr. Joshtrom Isaac Kureethadam), to the Buddhist call for "open hearts, open minds" from the noted spiritual teacher Jack Kornfield. Some authors are influenced by Jewish and Hindu traditions; some are secular and Marxist; and one (Michael Pollan) explores the role of psychedelics for helping us change our minds.

Several dozen academic disciplines are covered, and the authors represent every continent on the planet save Antarctica. But what counts for the most, we think, are the specific suggestions of "what we're learning about what's now possible." Varshini Prakash, co-founder of the youth-led Sunrise Movement, describes the strategies that Sunrise has adopted to bring about significant movement toward environmental justice and the Green New Deal. The global movement toward "commoning" is represented here by some of its most famous advocates, and you'll find that their chapters interweave in fascinating ways: David Bollier on commoning as a new paradigm, John Restakis on co-operatives in their newest emerging forms, Ellen Brown on public or commons banks, and a host of others. And Helena Norberg-Hodge, fresh from organizing a global conference on the subject, crafts a powerful vision of "localizing" ourselves after the failure of economic globalization.

This volume rings with the sound of scholar-activists and realist-visionaries. Virtually all the authors are involved in *bringing about* what they're writing about. Indeed, in one case the chapter and the work of an organization have evolved simultaneously. Although Justin Rosenstein's chapter is written in his voice, he credits the entire team of One Project for the concrete proposals advanced in the chapter. Over the coming months One Project will begin to prototype and field test the specific ideas that they describe in this important chapter. No chasms between theory and practice open up in these pages!

It's no coincidence that the book ends with David Korten, the best-selling author of *Change the Story, Change the Future*. "The new possible" is a shared quest to tell a new story, one that allows us to live lightly and justly on this beautiful blue planet. It's about using disruption to see possibilities and empower change, leveraging crisis to highlight new opportunities for a better world. In short, it's a book suffused and shot through with hope. May it lead you to your own new possibles!

~

The editors gratefully acknowledge the work of four professional editors who worked tirelessly to make this collection possible: Audra Sim (https://www.giantsquidedits.com/), Adrienne Ross Scanlan (http://adrienne-ross-scanlan.com/), Aaron Moschitto, and James Rogers. (Any remaining grammatical errors are our responsibility, not theirs!)

Finally, two partner organizations carried the burden of conceiving, producing, and funding this project from beginning to end, and we thank them for their substantial support: the Institute for Ecological Civilization (EcoCiv.org), and One Project (OneProject.org).

Essayists
(in the order in which they appear)

Kim Stanley Robinson (Preface) is an American novelist, widely recognized as one of the foremost living writers of science fiction. Robinson began publishing novels in 1984, and has been a proud defender and advocate of science fiction as a genre; which he regards as one of the most powerful of all literary forms.

Jeremy Lent (Chapter 1) is an author whose writings investigate the patterns of thought that have led our civilization to its current crisis of sustainability. His recent book, *The Patterning Instinct: A Cultural History of Humanity's Search for Meaning*, explores the way humans have made meaning from the cosmos from hunter-gatherer times to the present day. He is founder of the nonprofit Liology Institute, dedicated to fostering an integrated worldview that could enable humanity to thrive sustainably on the earth. His upcoming book is *The Web of Meaning: Integrating Science and Traditional Wisdom to Find Our Place in the Universe*.

Varshini Prakash (Chapter 2) is the executive director and co-founder of Sunrise, a movement of young people working to stop climate change, take back our democracy from Big Oil, and elect leaders who will fight for our generation's health and wellbeing. Varshini's work has been featured in *The New York Times, New Yorker* magazine, *TeenVogue*, BBC, *The Washington Post* and more. She was recently named to the *Grist* Top 50 Fixers and the *Time* 100 Next list.

Atossa Soltani (Chapter 3) has been a global campaigner for tropical rainforests and indigenous rights, now going on three decades. She is founder and board president of Amazon Watch and served as the organization's first executive director for eighteen years. Currently Atossa is the director of global strategy for Amazon Sacred Headwaters Initiative, working to protect

74 million acres in the most biologically diverse ecosystem on Earth. She is the Hillary Institute 2013 Global Laureate for Climate Leadership and is a producer of *The Flow*, a feature-length documentary currently in production about learning from nature's genius.

Justin Rosenstein (Chapter 4) is a programmer and designer of software, organizations, cultures, and systems in service of love. He is also co-founder of Asana, whose mission is to help humanity thrive by enabling all teams to work together effortlessly. Previously he was the, co-inventor of Google Drive, Gchat, Facebook Pages, and the Facebook Like button. Justin is dedicated to enabling all of humanity to collaborate toward a thriving sustainable joyful world for everyone. He is the founder of One Project, a social venture accelerating collaboration toward global thriving. More information can be found at oneproject.org.

Dr. Mamphela Ramphele (Chapter 5) is a South African human rights activist, physician, social anthropologist, businesswoman, and transformation leader known for her activism in the struggle for liberation against racism and sexism in South Africa. She continues to lead transformative initiatives as a co-Founder of ReimagineSA and co-president of the Club of Rome.

Anisa Nanavati (Chapter 6) is a high school student that was born and raised in Tampa, Florida. She is currently attends Academy at the Lakes in rural Land O'Lakes and is the North American coordinator for Earth Uprising. Anisa's activism includes plans to educate members of her community about the threats of the climate crisis and promote understanding among all, regardless of party alignment.

Michael Pollan (Chapter 7) has been writing books and articles for over thirty years, focusing on the places where nature and culture intersect: on our plates, in our farms and gardens, and in our minds. In addition to being a five-time *New York Times* bestselling author, he is a journalist, activist, and the Lewis K. Chan Arts Lecturer and Professor of Practice of Non-Fiction at Harvard University and Professor of Journalism at the UC Berkeley Graduate School of Journalism.

Dr. Riane Eisler (Chapter 8) is a social systems scientist, cultural historian, and attorney whose research, writing, and speaking have transformed the lives of people worldwide. Dr. Eisler is president of the Center for Partnership Studies, dedicated to research and education, and editor-in-chief of the *Interdisciplinary Journal of Partnership Studies*. She has addressed the United Nations and keynotes conferences worldwide. Her books include

The Chalice and the Blade (now in 56 US printings and 27 foreign editions), *The Real Wealth of Nations*, and most recently *Nurturing Our Humanity: How Domination and Partnership Shape Our Brains, Lives, and Future* (co-authored with anthropologist Douglas Fry).

Tristan Harris (Chapter 9), called the "closest thing Silicon Valley has to a conscience" by the *Atlantic* magazine, spent three years as a Google design ethicist working on how to fix the attention economy and technology's asymmetric influence over the thoughts and actions of billions of people. He is now co-founder and president of the Center for Humane Technology, a non-profit whose mission is to re-align the technology with humanity.

David Bollier (Chapter 10) is coauthor, with Silke Helfrich, of *Free, Fair and Alive: The Insurgent Power of the Commons*. He is director of the Re-inventing the Commons Program at the Schumacher Center for a New Economics, and co-founder of the Commons Strategies Group. He blogs at Bollier.org and lives in Amherst, Massachusetts.

Ellen Brown (Chapter 11) is an attorney, chair of the Public Banking Institute, and author of thirteen books including *Web of Debt, The Public Bank Solution*, and *Banking on the People: Democratizing Money in the Digital Age*. She also co-hosts a radio program on PRN.FM called "It's Our Money." Her 300+ blog articles are posted at EllenBrown.com.

Mark Anielski (Chapter 12) is a Canadian well-being economist, author, and an international expert in measuring the happiness and wellbeing of communities and businesses. He is the author of the award-winning book *The Economics of Happiness: Building Genuine Wealth* and *An Economy of Well-being: Practical Tools for Building Genuine Wealth and Happiness*. He is a former adjunct professor in both the School of Business, University of Alberta and the Bainbridge. *Alberta Venture* magazine named him one of the fifty most influential people in Alberta, Canada, in 2008.

Jess Rimington (Chapter 13) is a next economy strategist, practitioner, and scholar focused on ethics and methodologies of emerging post-capitalisms. Her research and practice is grounded in historical analysis, accessible truth-telling, present-day prototyping, and imagination. She is the co-author of the forthcoming *Beloved Economies: Transforming How We Work*, and considers herself a microeconomic activist. Jess is a third generation, female small-business owner and lives with her partner in Atlanta. www.jessrimington.org; www.belovedeconomies.org.

Natalie Foster (Chapter 14) is the co-chair and co-founder of the Economic Security Project, a network dedicated to a guaranteed income that would provide an income floor for families in America and anti-monopoly action to rein in the unprecedented concentration of corporate power. She's a senior fellow at the Aspen Institute Future of Work Initiative and Institute for the Future, and former director at Obama's Organizing for America, MoveOn.org and Sierra Club.

John Restakis (Chapter 15) has a master's degree in philosophy of religion and has been active in the co-operative movement for over twenty-five years. He is the former Executive Director of the Community Evolution Foundation and the British Columbia Co-operative Association. John is co-founder of Synergia Co-operative Institute, does consulting work on international co-op and community economic development projects, researches and teaches on co-operative economies and the social economy, and lectures widely on the subject of globalization, regional development, and co-operative systems change. He is the author of *Humanizing the Economy: Co-operatives in the Age of Capital.*

Dr. Vandana Shiva (Chapter 16) is trained as a physicist and in 2010 was identified by *Forbes* magazine as one of the top Seven Most Powerful Women on the Globe. She is the founder of Bija Vidyapeeth, an international college for sustainable living, and of Navdanya, a national movement to protect the diversity and integrity of living resources, especially native seed, the promotion of organic farming and fair trade. Among other distinctions, Dr. Shiva was identified by *Time* magazine as an environmental "hero" and *Asia Week* has called her one of the five most powerful communicators in Asia.

Dr. Mike Joy (Chapter 17) is a freshwater ecologist, science communicator, and Senior Research Fellow at the Institute for Governance and Policy Studies at Victoria University of Wellington, New Zealand. After seeing first-hand the decline in freshwater health in New Zealand, he became an outspoken advocate for environmental protection. Mike has received a number of awards for this work, including the inaugural New Zealand Universities Critic and Conscience Award (biennial, 2016). He was also a semi-finalist for the 2018 Kiwibank New Zealander of the Year and the 2013 Royal Society of New Zealand's triennial Charles Fleming Award recipient for Environmental Achievement. Dr. Joy works tirelessly to address the multiple environmental issues facing New Zealand and is passionate about sustainable food production.

Dr. Eileen Crist (Chapter 18) taught in the Department of Science, Technology, and Society at Virginia Tech for twenty-two years, retiring in 2020. She is the co-editor of a number of books including *Gaia in Turmoil: Climate Change, Biodepletion, and Earth Ethics in an Age of Crisis*; *Life on the Brink: Environmentalists Confront Overpopulation*; and *Keeping the Wild: Against the Domestication of Earth*. She has written numerous academic papers as well popular writings and is associate editor of the ecocentric online journal *The Ecological Citizen*. Her most recent book, *Abundant Earth: Toward an Ecological Civilization*, was published by University of Chicago Press in 2019. For more information and publications, visit her website www.eileencrist.com.

Dr. Zak Stein (Chapter 19) is a scholar at the Ronin Institute, where he researches the relations between education, human development, and the evolution of civilizations. He also serves as co-president and academic director of the activist think-tank at the Center for Integral Wisdom, where he writes and teaches at the edges of integral meta-theory. He has published two books, including *Education in a Time between Worlds*, along with dozens of articles. Zak is a long time meditator, musician, and caregiver, which has shaped him more than any professional engagements.

Oren Slozberg (Chapter 20) is the executive director of Commonweal in Bolinas, an organization dedicated to resilience, healing and justice with over twenty-five programs in Health and Healing, Environment and Justice and Education and Arts. He is also the program director of the Center for Creative Community at Commonweal which explores the intersection of dialogue, cognition, creativity, and community through summer camps, retreats, and community gatherings. Slozberg has been a senior program developer in the fields of education, youth development, and the arts for more than thirty years. Prior to Commonweal, Slozberg was a senior trainer with Visual Thinking Strategies in public schools and fine art museums in the USA.

Dr. Jack Kornfield (Chapter 21) trained as a Buddhist monk in the monasteries of Thailand, India, and Burma. He has taught meditation internationally since 1974 and is one of the key teachers to introduce Buddhist mindfulness practice to the West. After graduating from Dartmouth College in Asian Studies in 1967 he joined the Peace Corps and worked on tropical medicine teams in the Mekong River valley. He studied as a monk under the Buddhist masters Ven. Ajahn Chah, and the Ven. Mahasi Sayadaw Returning to the US, Jack co-founded the Insight Meditation Society in Barre, Massachusetts, with fellow meditation teachers Sharon Salzberg and Joseph Goldstein and the Spirit Rock Center in Woodacre, California. Jack has taught in centers and universities worldwide, led International Buddhist Teacher meetings, and worked with many of the great teachers of our time. His sixteen books have been translated into twenty languages and sold more than 1 million copies. He holds a PhD in clinical psychology and is a father, husband, and activist.

Dr. Francis Deng (Chapter 22) is currently Deputy Rapporteur of South Sudan National Dialogue and Roving Ambassador. He formerly held the positions of Sudan's ambassador to the Nordic countries, Canada, and the United States; Sudan's Minister of State for Foreign Affairs; the first Permanent Representative of South Sudan to the United Nations; Human Rights Officer in the UN Secretariat; Special Representative of the UN Secretary General on Internally Displaced Persons; and Special Advisor of the Secretary General for the Prevention of Genocide. He holds an Ll.B (honors) from Khartoum University, and Ll.M and J.S.D. from Yale University. He has authored or edited over forty scholarly books on a wide variety of subjects and two novels on the crisis of identity in the Sudan. Dr. Deng has held senior positions in leading American Universities and think tanks.

Fr. Joshtrom Isaac Kureethadam (Chapter 23) is coordinator of the Sector on "Ecology and Creation," Dicastery for Promoting Integral Human Development. He is also chair of Philosophy of Science and director of the Institute of Social and Political Sciences at the Salesian Pontifical University in Rome. He obtained a doctorate in philosophy at the Pontifical Gregorian University in Rome and was a Research Scholar to Campion Hall, University of Oxford, UK, where he is currently an academic visitor. His publications include *Creation in Crisis: Science, Ethics, Theology, The Philosophical Roots of the Ecological Crisis: Descartes and the Modern Worldview*, and *The Ten Green Commandments of Laudato Si'*.

Helena Norberg-Hodge (Chapter 24) Helena Norberg-Hodge (Chapter 24) is a pioneer of the new economy movement, and has been promoting an economics of personal, social and ecological well-being for more than thirty years. She is a recipient of the Alternative Nobel Prize, the Arthur Morgan Award and the Goi Peace Prize. Author of the inspirational classic *Ancient Futures*, she is also producer of the award-winning documentary *The Economics of Happiness*. Helena is the founder and director of Local Futures and The International Alliance for Localisation.

Dr. Rebecca Kiddle (Chapter 25) is a Senior Lecturer in Urbanism at Te Herenga Waka—Victoria University of Wellington, New Zealand. Her research interests are in the nexus between people and physical space, and the use of urban design as a tool to support the making of cities that work for both people and the environment. She is interested in Māori identity as it relates to towns and cities, and the role of young people in decision-making processes regarding the built and natural environment.

Dr. Arturo Escobar (Chapter 26) is an activist-researcher from Cali, Colombia. He was professor of anthropology and political ecology at the University of North Carolina, Chapel Hill, until 2018, and is currently affiliated with PhD programs in Design and Creation and in Environmental Sciences in Colombia. Over the past twenty-five years, he has worked closely with several Afro-Colombian, environmental and feminist organizations on ecological and transition issues. His best-known book is *Encountering Development: The Making and Unmaking of the Third World,* His most recent books are: *Designs for the Pluriverse: Radical Interdependence, Autonomy, and the Making of Worlds,* and *Pluriversal Politics: The Real and the Possible.*

Dr. David C. Korten (Chapter 27) holds MBA and PhD degrees from the Stanford Business School. He is a former member of the faculties of the Harvard Graduate School of Business and the Harvard Graduate School of Public Health. For twenty-one years, he lived and worked as an economic development professional in Ethiopia, Central America, the Philippines, and Indonesia and is founder and president of the Living Economics Forum. His books include *Change the Story, Change the Future: A Living Economy for a Living Earth,* and the international bestsellers *When Corporations Rule the World* and *The Great Turning: From Empire to Earth Community.*

Artists
(in the order in which they appear)

Marko Oblak (Earth) is a freelance artist and illustrator out of Seattle, Washington. As an avid outdoorsman, he draws his artistic inspiration from the natural environment, specifically through his passion for fly fishing. Many of his pieces are a representation of a backcountry adventure. As a self-taught artist, Marko specializes in intricate fine-line ink work and digital rendering. His work ranges from murals, custom tattoos, to print work. You can find more of his creations at www.creaturegenius.com.

Nikki McClure (Us) is an artist who lives in Olympia, Washington, under cedar trees along the eastern shore of the Southern Salish Sea. She cuts her images from single sheets of black paper with an X-Acto knife. She collaborates with her husband, Jay T. Scott, a fine woodworker, on lamps and furniture as well as daily walks. Her son is also a frequent collaborator and instigator of adventure as well as inspiration for many of her books and images. You can find more of her work at https://nikkimcclure.com.

Zaria Forman (Change) documents climate change with pastel drawings. She travels to remote regions of the world to collect images and inspiration for her work, which is exhibited worldwide. She has flown with NASA on several Operation IceBridge missions over Antarctica, Greenland, and Arctic Canada. She was featured on CBS *Sunday Morning*, CNN, PBS, and BBC. She delivered a TEDTalk; has spoken at Amazon, Google, and NASA's Goddard Space Flight Center; exhibited in Banksy's Dismaland; and was the artist-in-residence aboard the National Geographic Explorer in Antarctica. Her works have appeared in publications such as *The New York Times, National Geographic, The Wall Street Journal*, and *Smithsonian* magazine. Forman currently works and resides in Upstate New York, and is represented by Winston Wächter Fine Art in New York, New York, and Seattle, Washington. More of Zaria's work can be found at https://www.zariaforman.com.

Sam Wallman (Wealth) is a comics-journalist, cartoonist, and organizer based in Melbourne, Australia, and a member of the Workers' Art Collective. The piece included here was originally created for the Sunrise Movement, a movement of young people working to stop climate change, take back our democracy from Big Oil, and elect leaders who will fight for our generation's health and wellbeing founded by the author of Chapter 3, Varshini Prakash. Sam is a committed unionist, having worked as an organizer for the National Union of Workers, and a delegate on the shop floor prior to that. Three of his pieces of long-form comics-journalism have been nominated for Walkley Journalism Awards, including *Winding up the Window: The End of the Australian Auto Industry* and *A Guard's Story: At Work in Our Detention Centres*, which won the 2014 Australian Human Rights Award in the Print and Online Media category. More of his work can be found at https://www.samwallman.com.

Abby Paffrath (Work) was born and raised in Breckenridge, Colorado, and now lives in the Teton Valley of Wyoming. A true mountain girl, the outdoors have always been a huge part of her life. She has a bachelor's degree in Fine Art from the University of Montana and a master's in art education from Lewis and Clark College. Abby learned the art of batik while studying in Bali, Indonesia. She creates to slow down and connect with her environment, with the purpose of bringing joy to all those who wear and enjoy it. More of Abby's work can be found at https://art4allbyabby.com.

Lindsay Jane Ternes (Food) is an award-winning artist based in Boulder, Colorado. Born in Billings, Montana, and raised in Colorado Springs, Colorado, Lindsay's passion for art was apparent from a young age and never waned. At the University of Denver she majored in Marketing and Studio Art to equip herself for a future in fine art, and she rediscovered her love for landscape painting while studying at The Marchutz School of Fine Arts in Aix-en-Provence. Her time in southern France greatly influences her scene selection and light interpretation today. She typically works on location, painting *en plein air* around Colorado, but you'll also find her painting in Glacier National Park each summer. Lindsay's work and galleries can be found at https://www.lindsayjaneternes.com.

Favianna Rodriguez (Education and Cover) is an interdisciplinary artist, cultural strategist, and social justice activist based in Oakland, California. Her art and praxis address migration, gender justice, climate change, racial equity, and sexual freedom. Her practice boldly reshapes the myths, stories, and cultural practices of the present, while healing from the wounds of the past. More of Favianna's work can be found at https://favianna.com.

Mira Sachs (Love) is a 7th grader at Piedmont Elementary School in California with a passion for art, justice, and the natural world.

Lavie Raven (Community) is a social studies and language arts instructor at Oak Park and River Forest High School and serves as the Prime-Minister of Education for the University of Hip-Hop. Raven taught in the Chicago public school system for twenty years and has done community arts work since he was a teenager. In his artistic and pedagogical practice, he has created strategies for integrating hip-hop into community service projects and classroom education. He was the recipient of the Fund for Teachers Award, which he used to do hip-hop community work in British Columbia. Raven also received the Fulbright Distinguished Teacher Award and worked with Māori hip-hop activists and created hip-hop workshops for youth at schools in New Zealand.

Nina Montenegro (Tomorrow) is a Chilean-American visual artist, illustrator, and designer, whose artwork is in service to reconciliation and restoration. She is co-founder and co-creative director of the design studio The Far Woods with her sister Sonya. In March 2020, they released *Mending Life: A Handbook for Repairing Clothes and Hearts* with Sasquatch Books. More of her work can be found at https://www.thefarwoods.com and https://www.ninamontenegro.com.

Editors

Philip Clayton, Ingraham Professor at CST at Willamette University, holds a PhD from Yale University and has published several dozen books and some 200 research articles, as well as holding guest professorships at Harvard University, Cambridge University, and the University of Munich. Philip works at the intersection of science, religion, and ethics, and researches societal changes that are necessary for establishing sustainable forms of civilization on this planet. Philip is the president of the Institute for Ecological Civilization.

Kelli M Archie is Senior Science Advisor at the Institute for Ecological Civilization and Research Fellow in the Climate Change Research Institute at Victoria University of Wellington, New Zealand. She has a Ph.D. in Environmental Studies from the University of Colorado, Boulder and is a native of Colorado. Her previous research addresses climate change adaptation in the United States, the Indian Himalayas, Vanuatu and New Zealand. Kelli currently lives on the side of a ski run high in the Rocky Mountains with her husband and their four young daughters.

Jonah Sachs is the director of One Project. He is the co-founder of Free Range Studios, one of the minds behind the Story of Stuff series and the best-selling author of Winning the Story Wars.

Evan Steiner works to address global issues at their roots, focusing on structural issues within economics, finance, and business that are causing harm to people and the planet. His passion is building infrastructure that enables new economic paradigms based on ethics, ecological regeneration, and human flourishing. His current role is supporting partnerships and strategy for One Project, a social venture working to create new forms of governance and economics that are equitable, ecological, and effective.

EARTH

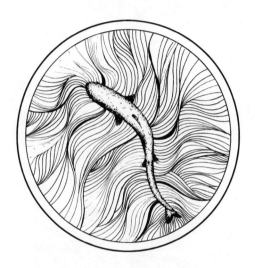

Trout Creek **by Marko Oblak**

1

Envisioning an Ecological Civilization

Jeremy Lent

DO YOU FEEL LIKE our society is coming apart at the seams? Does it seem like, as soon as one crisis passes, another one rears its head before you can even settle back to some semblance of normal? If so, you're not alone. Across the globe, people are beginning to realize that the world they inherited from previous generations is unraveling—and it's entirely unclear what will take its place.

When COVID-19 first swept across the globe in early 2020, there was a general sense that, if we all battened down the hatches, the storm would pass. We would go back to normal. But what is normal? For Black Americans, normal means not feeling safe in their own country. It means knowing that every day of their lives is a struggle against institutionalized racism. And it means that white Americans—even those who claimed to care—are doing nothing to change it. When George Floyd's brutal murder was caught on film, it kicked off a wave of protest across the country, waking up white people to their own silent complicity with an indelible message: there is no going back to normal—normal is unacceptable.

It's no coincidence that these waves of protests should occur in the middle of a pandemic that impacts Black and other minorities far more lethally than affluent white communities. And we can expect further disruptions over the coming months and years catalyzed by COVID-19. This is because the pandemic reveals the structural faults of a system that have

3

been papered over for decades. Gaping economic inequalities, rampant ecological destruction, and pervasive political corruption are all results of imbalanced systems relying on each other to remain precariously poised. As one system destabilizes, expect others to tumble down in tandem in a cascade known by researchers as "synchronous failure."

Ultimately, there is no going back to normal because normal no longer exists—except in the messages of the mainstream media and politicians paid to keep the public in a consensus trance while a small elite sucks the wealth out of human communities and natural ecosystems, all in the name of the dominant ideology of neoliberalism.

Neoliberalism has been a part of global mainstream discourse since the 1980s. It propagates the fiction that humans are essentially individualistic, selfish, calculating materialists, and as a result, unrestrained, free-market capitalism provides the best framework for every kind of human endeavor. Through their control of government, finance, business, and media, neoliberal adherents have transformed the world into a globalized market-based system, loosening regulatory controls, weakening social safety nets, reducing taxes, and virtually demolishing the power of organized labor.

Neoliberalism is the logical outcome of a worldview based on separation: people are separate from each other; humans are separate from nature; and nature itself is no more than an economic resource. The value system built on this foundation is the ultimate cause of the world's gaping inequalities, our roller-coaster global financial system, our failure to respond to climate breakdown, and our unsustainable frenzy of consumption.

Heading for Collapse?

As our civilization hurtles toward the precipice, we will encounter ever greater shocks that will make COVID-19 seem like a leisurely rehearsal by comparison. This is a result of our society's ecological overshoot: the fact that, in the pursuit of material progress, we are consuming the Earth's resources much faster than they can be replenished. Our civilization is currently running at forty percent above its sustainable capacity, with no plans to reduce growth in consumption. Imagine you had a friend who inherited vast wealth but could only access it in the form of an annuity. Dissatisfied with his annual stipend, he hires lots of lawyers and accountants to find ways to retrieve his wealth more rapidly. Now, living in palatial surroundings, his accountants warn him that he will soon clean out his account, but he ignores them. If he then turns to you and proudly claims that his opulent lifestyle is a sign of progress, you might consider him to have lost his mind. This is the current state of our global society.

While there are multiple dangers arising from our society's reckless overshoot, the looming climate catastrophe is without doubt the most dire. Since the UN's Intergovernmental Panel on Climate Change (IPCC) issued a warning to humanity in 2018 that we have just twelve years to turn things around before passing the point of no return, the world has continued its trajectory of uncontrolled emissions. Scientists warn of tipping points in the climate system that, once triggered, cause a cascade of further tipping points, leading to runaway climate catastrophe—and an unrecognizable world utterly incompatible with any form of civilization. It's possible we've already passed the point of no return—no one knows for sure—but we know that each year of growth in emissions brings us closer to that doomsday point.

Even if the climate crisis were somehow brought under control, continued economic growth in future decades will bring us face-to-face with a slew of further existential threats. We're rapidly depleting the Earth's forests, animals, insects, fish, freshwater—even the topsoil we need to grow our crops. We've already transgressed three of the nine planetary boundaries that define humanity's safe operating space. Global gross domestic product (GDP) is expected to triple by 2060, with potentially irreversible and devastating consequences. In 2017, over 15,000 scientists from 184 countries issued an ominous warning to humanity that we are running out of time to shift away from our failing trajectory—it will soon be too late.

A Transformation in Values

As long as government policies emphasize GDP growth and transnational corporations relentlessly pursue shareholder returns, we will continue accelerating toward global catastrophe. These practices ransack the Earth without regard to long-term effects. If we are truly to "shift course away from our failing trajectory," the new era must be defined, at its deepest level, not merely by the political or economic choices we make, but by a transformation in the very way we make sense of the world, and by a concomitant revolution in our predominant values.

The depiction of humans as selfish individuals, the view of nature as a resource to be exploited, and the idea that technology alone can fix our biggest problems are all profound misconceptions that have collectively led our civilization down this madcap path to disaster. We must recognize the destructive nature of the dominant mainstream culture and reject it for one that is life-affirming, embracing values that emphasize growth in the quality of life rather than in the consumption of goods and services. We must emphasize core human values of fairness, justice, and compassion as paramount—extending them through local neighborhoods to state and national government, to the global community of humans, and ultimately to the community of all life.

In short, we need to change the basis of our global civilization. We must move from a civilization based on wealth accumulation to one based on the health of living systems: an *ecological civilization*. A change of such magnitude would be an epochal event. There have only been two occasions in history when radical dislocations led to a transformation of virtually every aspect of the human experience: the Agricultural Revolution that began about twelve thousand years ago, and the Scientific Revolution of the seventeenth century. If our civilization is to survive and prosper through the looming crises of this century, we will need a transformation of our values, goals, and collective behavior on a similar scale.

An Ecological Civilization

An ecological civilization would be based on core principles that sustain living systems in natural ecologies. Over billions of years on Earth, life has evolved resilient processes that allowed it to spread in rich profusion and stunning diversity into virtually every nook and cranny of the planet. As a result, if left undisturbed by human depredation, natural ecosystems can persist in good health for millions of years.

Living systems are characterized by both competition and cooperation. However, the major evolutionary transitions that brought life to its current state of abundance were all the results of dramatic increases in cooperation. The key to each of these evolutionary steps—and to the effective functioning of all ecosystems—is *symbiosis*: the process by which both parties in a relationship give and receive reciprocally, reflecting each other's abilities and needs. With symbiosis, there is no zero-sum game; the contributions of each party create a whole that is greater than the sum of its parts.

An important result of symbiosis is that ecosystems can sustain themselves almost indefinitely. Energy from the sun flows seamlessly to all the constituent parts. The waste of one organism becomes the sustenance of another. In contrast to our current civilization, which built its wealth by extracting resources and letting waste accumulate, nature has a *circular economy* where nothing is squandered.

The complex interconnection of different organisms in a symbiotic network leads to another foundational principle of nature: *harmony*.

Harmony doesn't mean bland agreement. On the contrary, it arises when different elements within a system express their own needs so that the system as a whole is enriched. Harmony arises when the various forces of the system are in balance. This can manifest as balance between competition and cooperation; between the system's efficiency and its resilience; or between growth, maturation, and decline.

In a natural ecosystem, the type of exponential growth that our civilization is currently experiencing could only occur if other variables were already out of balance. Growth of that sort would lead inevitably to the catastrophic collapse of that population.

From microscopic, intracellular structures to the entire Earth system, nature uses a fractal design. Coherent self-organized entities are embedded within larger systems: cells are part of an organism, which is part of a population, which is part of an ecosystem. In all cases, the health of the system as a whole requires the flourishing of each part. Each system is interdependent on the vitality of all the other systems. This universal precept leads to the ultimate objective of an ecological civilization: to create the conditions for all humans to flourish as part of a thriving, living Earth.

The symbiosis intrinsic to natural systems translates in human terms into foundational principles of *fairness* and *justice*, ensuring that the efforts and skills people contribute to society are rewarded equitably. Recognizing that the flourishing of the system as a whole requires the health of all its constituent parts, an ecological civilization would foster *individual dignity*, providing the conditions for everyone to live in safety and comfort, with universal access to proper housing, competent healthcare, and quality education.

An ecological civilization would celebrate *diversity*, recognizing that its overall health depended on different groups—self-defined by ethnicity, gender, or any other delineation—developing their own unique gifts to the greatest extent possible. It would be built on the axiom that a system's full potential can only be realized when it is truly *integrated*—a state of unity with differentiation, wherein the flourishing of each constituent part generates the wellbeing of the larger whole.

The principle of *balance* would be as crucial to an ecological civilization as it is to natural systems. Competition would be balanced by collaboration. Disparities in income and wealth would remain within much narrower bands and would fairly reflect the contributions people make to society. In the same way that an animal requires robust circulation of its lifeblood to maintain health, a life-affirming economy would be designed to enable the widespread circulation of its wealth throughout the entire community. And crucially, growth would become part of a natural life cycle, slowing down to maturation once it reaches its healthy limits—leading to a steady-state, self-sustaining economy designed for wellbeing rather than consumption.

Above all, an ecological civilization would give rise to an all-encompassing *symbiosis* between human society and the natural world. Human activity would be organized, not merely to avoid harm to the living Earth, but to actively regenerate and sustain its health into the long-term future.

An Ecological Civilization in Practice

Transitioning to an ecological civilization would require fundamentally redesigning our economy. Across the world, the success of political leaders is currently measured by how much they've managed to increase their nation's GDP. However, GDP merely measures the rate at which a society is transforming nature and human activities into the monetary economy, regardless of the ensuing quality of life. Anything that causes economic activity of any kind, whether good or bad, adds to GDP. In place of an economy based on perpetual growth in GDP, a life-affirming society would emphasize growth in quality of life, using alternative measures such as the "Gross National Happiness" index established by the state of Bhutan, which assesses qualities such as spiritual wellbeing, health, and biodiversity.

Ever since the nineteenth century, most economic thinkers have recognized only two domains of economic activity: markets and government. The great political divide between capitalism and communism arose from stressing one or the other of these two poles (with social democracy somewhere in between). An ecological civilization would incorporate government spending and markets, but—as laid out by progressive economist Kate Raworth—would add two critical realms to the old framework: households and the commons.

In particular, the commons would become a central part of economic activity. Historically, the commons referred specifically to shared land that peasants accessed to graze their livestock or grow crops. But in a broader context, the commons refers to *any* source of sustenance and wellbeing that is not appropriated either by the state or private ownership: the air, water, sunshine, and even human creations like language, cultural traditions, and scientific knowledge. The commons is virtually ignored in most economic discussions because, like household work, it doesn't fit into the classic model of the economy. But the global commons belongs to all of us. In an ecological civilization, it would once again take its rightful place as a major provider for human welfare.

The cumulative common resources that our ancestors have bequeathed to us through untold generations of hard work and ingenuity represent a vast reservoir of wealth—our shared human *commonwealth*—compared to which the value added by any individual is a drop in the ocean. An ecological civilization, recognizing this, would fairly reward entrepreneurial activity but would severely curtail the right of anyone to accumulate multiple billions of dollars in wealth, no matter what their accomplishments.

On the other hand, it would recognize the moral birthright of every human to share in this vast commonwealth. The transition could effectively be achieved through a program of unconditional cash disbursements to every person alive on the planet, known as *universal basic income*. The dominant neoliberal view of human nature leads many to assume that free money would cause people to become lazy, avoid work, and exacerbate addictive behaviors. In every test conducted, however, the opposite turns out to be true. Programs consistently report reductions in crime, child mortality, malnutrition, truancy, teenage pregnancy, and alcohol consumption, along with increases in health, gender equality, school performance—and even entrepreneurial activity. For these moral and practical reasons, universal basic income would be integral to the design of an ecological civilization.

The transnational corporations that currently dominate virtually every aspect of our global society would be fundamentally reorganized and made accountable to the communities they purportedly serve. Corporations above a certain size would be required to be chartered with the explicit purpose of optimizing not just for shareholder returns, but also for social and environmental outcomes. This approach—sometimes referred to as the "triple bottom line" of people, planet, and profits—already exists in the form of what are known as certified B-corps and benefit corporations. Currently, these charters are voluntary, and very few large corporations adopt them. If, however, they were compulsory, it would immediately transform the intrinsic character of corporations. Strict enforcement procedures would be necessary to ensure all three bottom lines were optimized. In my proposal, these would include the threat of a corporation losing its charter to do business, based on regular determinations by panels composed of representatives of the communities and ecosystems that are covered in the company's scope of operations.

In place of the vast homogenized monocropping of industrial agriculture, food would be grown worldwide based on principles of regenerative agriculture, which means greater crop biodiversity, improved water and carbon efficiency, and the virtual elimination of synthetic fertilizer. Manufacturing would prioritize circular flows with efficient re-use of waste products built into processes from the outset. Locally owned cooperatives would become the default organizational structure. Technological innovation would still be encouraged, but would be prized for its effectiveness in enhancing the vitality of living systems rather than minting billionaires.

Cities would be redesigned on ecological principles, with community gardens on every available piece of land, essential services always available within a twenty-minute walk, and cars banned from city centers. The local community would be the basic building block of society, and face-to-face interactions would again become a central part of human flourishing. Education would be re-envisioned, its goal transformed from preparing students for the corporate marketplace to cultivating the wisdom, discernment, and emotional maturity that are required for each student to embark on a lifetime of pursuing their own wellbeing as valued members of society.

Local community life would be enriched by the global reach of the internet. Online networks that have already achieved scale, such as Facebook, would be turned over to the commons, so that rather than manipulating users to maximize advertising dollars, the internet could become primarily a vehicle for humanity to further develop a planetary consciousness. Cosmopolitanism—an ancient concept from the Greeks which literally means "being a citizen of the world"—would be the defining character of a global identity. We would celebrate diversity between cultures while recognizing the deep interdependence that binds all people into a single moral community with a shared destiny.

Governance would be transformed to a polycentric model, where local, regional, and global decisions are made at the levels where their effects are felt most. While most decision-making would devolve to the lowest feasible level, a stronger global governance would enforce rules on planetary-wide issues, such as contending with the climate emergency and halting the Sixth Extinction. A worldwide Rights of Nature declaration would put the natural world on the same legal standing as humanity. Personhood would be ascribed to ecosystems and high-functioning mammals, and the crime of ecocide—the destruction of ecosystems—would be prosecuted by a court with global jurisdiction.

Toward the Great Transition

While this vision may seem a distant dream to those who are transfixed by the daily frenzy of current events, innumerable pioneering organizations around the world are already planting the seeds for a life-affirming civilization.

In the United States, the visionary Climate Justice Alliance has laid out the principles for a just transition from an extractive to a regenerative economy that incorporates deep democracy with ecological and social wellbeing. In Bolivia and Ecuador, traditional ecological principles of *buen vivir* and *sumak kawsay* ("good living") are written into the constitution. In Europe, large-scale thriving cooperatives, such as Mondragon in Spain, demonstrate that it's possible for companies to provide effectively for human needs without utilizing a shareholder-based profit model.

Meanwhile, a new ecological worldview is spreading globally throughout cultural and religious institutions, establishing common ground with Indigenous traditions that have sustained their knowledge and lifeways for millennia. The core principles of an ecological civilization have already been set out in the Earth Charter—an ethical framework launched in The Hague in 2000 and endorsed by over two thousand organizations worldwide, including many governments. In China, leading thinkers espouse a New Confucianism, calling for a cosmopolitan, planetary-wide ecological approach to reintegrate humanity with nature. In 2015, Pope Francis shook the Catholic establishment by issuing his encyclical, *Laudato Si'*, a masterpiece of ecological philosophy that demonstrates the deep interconnectedness of all life and calls for a rejection of the individualist, neoliberal paradigm.

On the secular front, economists, scientists, and policymakers, recognizing the moral bankruptcy of the current economic model, are pooling their resources to develop alternative frameworks. The Wellbeing Economy Alliance is an international collaboration of changemakers working to transform the present global economic system to one that promotes human and ecological wellbeing. Similarly, the Global Commons Alliance is developing an international platform for regenerating the Earth's natural systems. Organizations such as the Next System Project and the Global Citizens Initiative are laying down parameters for the political, economic, and social dimensions of an ecological civilization, and the P2P Foundation is building a commons-based infrastructure for societal change. Around the world, "transition towns" are growing into an international movement and transforming their communities from the grassroots up.

Perhaps most importantly, a people's movement for life-affirming change is spreading globally. When Greta Thunberg skipped school in August 2018 and went to the Swedish parliament in order to draw attention to the climate emergency, she sat alone for days. Less than a year later, over one and one-half million schoolchildren joined her in a worldwide protest to rouse their parents' generation from their slumber. A month after Extinction Rebellion demonstrators closed down Central London in April 2019 to draw attention to the world's dire plight, the UK Parliament announced a "climate emergency." Similar declarations have now been made by over 1,500 local and national governments worldwide, representing over ten percent of the global population. Meanwhile, a growing campaign of "Earth Protectors" is working to establish ecocide as a crime prosecutable by the International Criminal Court in The Hague.

Is this enough? When we consider the immensity of the transformation needed, the odds look daunting. Yet no one can say that the task is impossible. As the world system begins to unravel on account of its internal failings, the strands that kept the old system tightly interconnected also get loosened. Each year that we draw closer to catastrophe—as greater climate-related disasters rear up, as the outrages of racial and economic injustice become even more egregious, and as life for most people becomes increasingly intolerable—the old story loses more and more of its hold on the collective consciousness of humanity. Waves of young people are looking for a new worldview—one that makes sense of the current unraveling, one that offers them a future they can believe in. Remember that people who lived through the Industrial Revolution had no name for the changes they were undergoing; it would be a century before their age received its title. Perhaps the Great Transition to an ecological civilization is already occurring, but we can't yet see it fully because we're in the middle of it.

The chapters of this book that follow help to blaze the trail for what is now possible. As you read them, I invite you to consider what brings the most excitement into your heart, and then to explore that direction further in your own engagement. Like an immune system protecting its host from toxins, more and more caring, compassionate humans are drawing together in expanding networks and devoting their energies to life-affirming activities. An ecological civilization will only emerge when enough people around the world decide they no longer want to allow humanity to hurtle off the precipice and begin to work together collaboratively to shift the direction of our species. Each of us has a part to play in co-creating our destiny and bequeathing a vibrant Earth to future generations.

2

Speaking Our Needs for the Future

VARSHINI PRAKASH

Adapted from an interview conducted on July 31, 2020

WE ARE IN A unique moment. We are in the midst of a pandemic that has rendered millions of people jobless and killed over 150,000 people in the United States. The pandemic has exposed the shortcomings of a profit-motivated health-care system. Simmering crises around police brutality, racism, and white supremacy have worsened. We are seeing one of the hottest climate years on record. The 2020 Atlantic hurricane season has started months early, with storms already crashing ashore.

The crises of our generation have reached a fever pitch. Young people are fed up with the status quo. They are fed up with the way that our systems and institutions have been operating, and they now recognize whom they have been serving—some of them since the dawn of America. People young and older are willing and ready to risk their lives in order to secure a more just future and reality for themselves and their loved ones. The moment is terrifying, and yet it is also bright with opportunity.

We are young people who have witnessed a world in chaos careening towards climate catastrophe. We have watched and waited our entire lives for people much older and more powerful than us to take care of the crises that were emerging. Yet little has happened. Now our generation is standing up to say, "We are ready to be the adults in the room. We are ready to take

13

the future into our own hands. We are ready to envision reality in a different way."

The Beginning of a Movement

A defining moment for me was December 2015 when a series of extremely strong floods deluged Tamil Nadu, the state in India that my dad (and a lot of my family) is from. It was amazing to me, looking at the images on my computer from half a world away, to be able to see the streets that I had walked on as a kid with my grandma or driven on with my grandfather in his little car. Suddenly I was seeing women and children who somehow looked very familiar to me walking waist deep or chest deep in water, traveling miles to sanctuary.

My grandparents were, fortunately, not in town at that time, but the water had come all the way up to their apartment floor. My grandmother told me how people had been stranded in the building with the power gone out. I remember thinking, "If my grandparents had gotten stuck there for days or weeks with no food, no water, no electricity, would they have survived? What would have had to happen for them to survive?" They were extremely lucky. But a lot of other people were not. Hundreds of people died in that flood, and thousands were displaced. That was 2015, and it was a big wakeup call to me that the climate crisis was right now. The increase in the number and severity of flooding episodes—predicted as a result of climate change—was happening now, in the present, not in the future.

That was the moment. I thought, *What do I have to lose?* This time it was someone else's grandmother; the next time, it could be mine. We didn't have time to waste. The movements that were prominent in that moment were not numerous enough, and they were not as powerful as we needed them to be. That very same month, my friend Sarah and I decided we were going to start an environmental movement for young people. We needed a movement that could be powerful and could grow quickly—quick enough to respond to the climate crisis as it is worsening all around us. This was the beginning of Sunrise.

The Power of the Young

What we found when we were creating Sunrise was that there was no political home for young people in America who were concerned about the climate crisis. There was no political home for teenagers and twenty-somethings who woke up every day horrified by the crisis and went to sleep imagining a chaotic, climate-disordered world. We realized that it would be absolutely game changing if we could harness the power of young people—all their passion, optimism, and hope—and translate it into campaigns for long-lasting political change.

Young people have historically played an important role in social movements and political change. John Lewis, who passed away a week before the writing of this chapter, was just a college student when he became a leader in some of the most intense protests of the civil rights movement. Then there was Diane Nash, whose youth campaigns were crucial to its success. James Lawson organized young people on campuses across the country in large part because he understood the role that young people played—in being willing to take risks and have courage, not living or being governed by the rules of yesterday. I think that's one of the most unique things about young people: we're not jaded about what is or isn't possible. We just know what needs to happen, and we work like hell to make sure that it does. That's how progress happens.

One of the biggest moments that has occurred in Sunrise's first two years came right after the 2018 midterm elections. Young people with Sunrise had campaigned for dozens of candidates around the country. We had knocked on thousands of doors, made tens of thousands of phone calls, and worked to elect climate champions to congressional and local offices. But based on past decades we knew that, even having taken back the House of Representatives, Democrats still would not prioritize the climate crisis—instead, they would wait until they regained control of the House, Senate, and presidency. And we knew that would be too late.

So, right after the 2018 midterms, two hundred Sunrisers marched to Nancy Pelosi's office (she was about to become Speaker of the House) and stood in the lobby with signs that said, "What is your plan?" We told stories, sang songs, and chanted (dozens of us were arrested), calling on Nancy Pelosi and the Democratic Party to back a solution appropriate to the scale of the climate crisis, a solution like the Green New Deal. We said that it wasn't enough just to believe in the science. We wanted them to say, "We must, as Democratic leaders, swear off oil and gas money moving forward." I remember one poignant moment where one of our leaders, Claire Tacherra-Morrison, told her story of how her aunt and uncle had lost their home in the California wildfire. At the very same moment, the TV in the lobby was showing images of the Camp Fire as it burned people's houses in California. Alexandria Ocasio-Cortez joined us halfway through and gave a rousing speech about why the activism of young people is so essential for pushing the Democratic Party to lead on the issue of climate in the ways that we so badly need and deserve.

This protest really catapulted climate change into the public consciousness. It completely transformed the conversation about climate change in the country. All of a sudden, politicians began reflecting back a conversation about climate that mirrored what frontline and environmental justice leaders had been saying for decades. We saw almost five thousand articles written about the Green New Deal in forty-eight hours. Ever since then, the Green New Deal has been a litmus test for serious politicians running for the highest offices in the country. It has gained bipartisan support and become one of the most recognizable markers of climate action and policy in America and around the world.

How to Work Like a Youth Movement

Youth movements have a particular approach to working for change, and we at Sunrise have been inspired by them. I think of four lessons in particular that we've learned from them. One of the biggest and most important principles of effective protest is this: in your demands and your vision, don't lead with what is *possible* in today's reality, but with what is *necessary*—for, say, the survival of humanity, or for achieving the ultimate goals of whatever campaign or issue you're working on. So often, I find that older generations are hindered by their view of what is possible or impossible right now. The most common thing I hear is "It's not practical" and "It's not realistic, considering who's in office right now" and so on. Everything that Sunrise has achieved has been under a Trump administration. If we had been led by that more pragmatic doctrine, if we had based our goals and our ambitions on the parameters of today's political possibilities, we would never have been successful.

Second, we were unafraid to go not just after Republicans—who were denying the validity of climate science and supporting misinformation campaigns—but also after Democrats. We said to them, "You, too, have not done enough on this issue. You have said you believe the science. You have voted the right way. But truly, we need champions. We need fighters. We need people who are ready to stick their necks out on issues, who will fight day in and day out, who will be the leaders that we so badly need on the critical issues, and talk about them from racial and economic justice perspectives." What's crucial is being willing to call for the level of action you want, no matter what your political affiliation is.

Third, storytelling is powerful. When we went into Nancy Pelosi's office, we didn't just deliver a petition with a bunch of numbers about parts per million or two degrees Celsius. We shared stories about what we had lost because of the climate crisis, or what we were afraid of losing. We told stories about what we hoped for our future. Some of the storytellers were in high school, not even able to vote yet, but were engaged in politics because of how much they cared for their future. People told stories about what it was like to live through hurricanes and come out the other end, about the trauma these experiences instill, and about their hope that such traumas don't have to be the story for future generations. Storytelling and narratives are essential, even in visual depictions. If you look at any image of the Nancy Pelosi protest, it's striking to see the signs staring back saying, "What is your plan?" in gray and gold. That imagery has been shared all over the world. To my mind, they are some of the most poignant images of the youth climate movement.

Fourth, young people are amazing these days at using all of the tools at our disposal to reach other young people, sharing our ideals not just from a political perspective, but also from a cultural perspective. We powerfully marry digital organizing with offline organizing, using humor, comedy, Tik-Tok, Instagram, and other tools. We saturate culture with our ideas, reaching people through song, art, video, and graphics. Many people have told me they joined Sunrise because they really liked our logo. We thought a lot about our logo and its meaning; we also had a designer work carefully on it and come back to us with multiple iterations. Your visuals and communication, both digitally and offline, have to communicate something significant to people.

Black Lives Matter and the Climate Crisis

The climate crisis and racial injustice are deeply related. Hurricane Katrina claimed three thousand lives—the vast majority of which were Black lives, elderly lives, and working people's lives—because our society had deemed some lives to be worth less than others. Regardless of whether Katrina can be attributed to climate change or not, the climate crisis has been created in large part because some people's lives are held to matter more than others. The Global North can pollute and burn because the people who end up suffering are those in the Global South—people of color, or working people who have far less political capital and economic capital, or people who are unable to say, "No, enough is enough."

So, the way I see it, it's not a question of how the issues of climate and racial justice bridge together, because they are not separate to begin with. Rather, they have the same causes, and they have the same impact. The causation is a system of pervasive inequality—a system of unfair, unjust inequality in this country. People have built a hierarchy of human value, willingly, for the sake of economic profit. We created an unequal society at the very foundation of this country's democracy, when millions of Indigenous people were murdered in a mass genocide, when they were routinely marginalized and assimilated, and when we propped up a system of chattel slavery that traded Black lives for economic profit.

When there are people who can be treated as non-human, it becomes possible to have things like sacrifice zones, where fracking wells are located within 150 feet of a school facility, and to have pipelines that so pollute lands and waterways that frogs are born with two heads and three legs. When I read calls to "defund the police," I hear Black people saying, "Stop investing in the things that kill us. Stop investing in a carceral system that inflicts pain and punishment. We don't need brutality and surveillance. We need good schools, good education, and jobs that pay well. We need childcare, clean air, and clean water. We need to not be pinched and squeezed at every single turn. Invest in Black dignity and life so that we can have lives and thrive."

When we talk about a Green New Deal, we're saying, "Stop investing in the things that kill us, that create toxins in our water, that cause asthma for our babies." Stop polluting. Invest instead in things that are good, like affordable and safe and clean housing, good jobs, clean air, clean water, and healthcare." The fights for racial justice and the climate have the same vision for the future that we collectively need. They are so deeply intertwined that if we fail to tackle both at the same time, I'm convinced that we will lose.

Politics and Organizing in the Pandemic

Right now, we are seeing an expansion of what is possible. Take for example someone like Joe Biden, the very definition of a moderate candidate, who, six months ago, had one of the weakest climate plans among all the Democratic candidates. The pandemic hit, and then a massive uprising around racial injustice took the country and the world by storm. Biden has defined himself over his career largely as an incrementalist. Yet now, because of these huge systems-disrupting problems and the calls for transformative change, he's being forced to consider far bigger, broader, and more transformational solutions. They might actually be systems-*shifting* reforms. For example, his climate plan went from being a $1.7 trillion green jobs and infrastructure plan over ten years to a $2 trillion plan over four years, with forty percent of those investments—$500 billion—going directly to frontline communities. It's hard to even fathom what that could do for communities of color and poor people around the nation. It's far more than any other president has committed on this issue.

Naomi Klein frequently talks about how, when there are moments of great crisis, opportunities emerge for a country to go in any number of possible ways. We have an opportunity now to say, "There are forty million people unemployed. Do we want to go back to the same unjust economy of the past, or do we want something better? Do we want to go back to the same system of militarized police and brutality that started with slavery? Or do we want to forge a different path for the United States?"

It's about recognizing choice points like these in many different sectors. Right now, we are seeing the possibility to make every job in the new economy a union job that has access to good wages and meets health standards. We can expand Indigenous rights to ensure that things like the Dakota Access Pipeline and the standoff between militarized police and Indigenous people protecting their land and water never have to happen again. All of this is becoming possible. That is truly remarkable. It's a testament to the crisis and to our organizing—for solutions we really need in the moment—that these choice points are coming together with the potential to lead to extremely transformative changes.

Still, organizing in the pandemic has been difficult. Doing this work online has been challenging for the Sunrise movement, which is so based in song and sharing stories and having real human connections. Other climate movements before us did not have that emotional, relational touch, which was why people did not stick around. Organizing online has felt inhibiting.

Nonetheless, we've managed to train almost fifteen thousand people virtually just in the last couple of months. We've mobilized about a dozen staff to support the Movement for Black Lives full time. We've supported our leaders in figuring out how they could take action in this moment. We've supported thousands of young people who are actively joining the movement in the streets, training them to participate in the process safely, responsibly, and respectably. We've partnered with racial and immigrant justice organizations to give a crash course on how police brutality and defunding the police are related to the Green New Deal, which was very well attended. As a result of these efforts, I am excited to say, the Movement for Black Lives has officially joined the Green New Deal Network, a broad and diverse coalition fighting for a Green New Deal at the federal and state level.

So, there are many ways of making a difference and continuing to work for social change during the pandemic. There have even been instances of individual Sunrise organizers mobilizing official government resources through their own example. A colleague of mine who lives in Birmingham, Alabama began actively sharing and distributing food and supplies in her community the moment that COVID-19 happened, working especially to alleviate the problems caused in the Black community by gentrification. The mayor of Birmingham, a powerful Black woman who had recently been elected to office, began following the efforts of my colleague's community group and eventually based a citywide policy on their approach. This was very exciting and powerful to see.

Building Communities for the Future

The biggest thing that needs to happen for a better future is that ordinary people need to get more power. I don't expect power holders or people in office to do this for us. We have to build movements. In particular, we need to rebuild youth movements and the labor movement. We have to have the discipline, strategic acumen, and intellect of the fighters who have come before us. And we have to grow our ranks by orders of magnitude.

The truth is that you can dream up all the white papers you want and create all the policy proposals you want, but we can't enact any of it into reality if we don't have power. That is the bottom line for me when answering almost any question about what is and isn't possible in the next few years. The road forward is uncertain. But the question of what's possible stretches us to open up our imagination and create new worlds in ways that we might never have dreamed of before.

3

The Pandemic
Portal to a Living Earth Paradigm

ATOSSA SOLTANI

IN EARLY MARCH 2020, just a handful of months after the worst fire season in decades ravaged the Amazon and Australia's forests, a highly contagious novel virus started attacking the lungs and hearts of humans across the globe. The parallels were hard to miss.

The past months of living through this global pandemic have given us a pause, a time to reflect on our present moment, a time to imagine a different future. There can be no going back to the "old normal" of a life-blind and predatory economic system that has pushed the Earth's life support systems to the brink. From my living room, on constant Zoom calls with activists and thinkers from around the globe, I have been going deep in exploring the meaning and implications of this time. We are in a liminal space. Though the future may be hazy, there are glimpses of it on the horizon.

This pandemic may be the most significant opportunity in our lifetime for advancing societal shifts towards a more just and harmonious human presence on Earth—that is, *if* we seize this moment. Central to this moment is working as allies and visionaries alongside Indigenous peoples to protect the Amazon rainforest for our planet's health and our common future.

The Pandemic and the Possible

"Another world is not only possible, she is on her way.
On a quiet day, I can hear her breathing."

—Arundhati Roy

A big lesson of the pandemic has been that we are capable of making swift and radical shifts in the face of crisis. Simply look at how unprecedented grassroots mutual-aid networks have begun to self-organize in communities in order to care for and support each other. When we are faced with an existential threat, we can choose life over money, and we can cooperate in creative ways.

We must bring this same compassionate creativity to the climate and biosphere emergencies that existed before the pandemic. As a climate and forest activist, I follow the constant stream of data about the myriad ecological crises we face. Global warming, rainforest destruction, mass extinction, soil depletion, ocean acidification, and desertification are all signs of the impending collapse of our biosphere's life support systems. The implications of our planetary crises will be even more catastrophic for humanity than the coronavirus. And let's not forget that the proliferation of novel viruses is often linked to ecological destruction.

This pandemic is offering us a momentary portal to reframe the dominant paradigm and thus to catalyze a corresponding shift in our cultural beliefs, our values, and our priorities. For one important example, consider how, just as political leaders started planning a return to normal, individuals and social movements around the world rose up in unprecedented ways to stand with Black Lives Matter. Millions took to the streets in the middle of a pandemic to demand justice and real change, risking their own health and safety to challenge systemic injustice and institutional racism.

The same mindsets, cultures, and economic systems that perpetuate oppression, domination, racism, and colonization also perpetuate violence and destruction of our Earth. We can and must challenge the paradigm that sees the planet and its life-giving rainforests as a storehouse of "resources" to be exploited in the quest for progress, "development," and wealth creation in what they view as a hostile competitive world. We can demonstrate scientifically that we are co-existing in a complex web of life, on a living planet that is more like an organism than a floating rock in space, a planet on which communities of life self-organize and create conditions conducive to more life.

Yet we have very little time. Let me repeat that: *we have very little time.* The UN says we have at best twelve years to address the crises of climate and mass extinction. Incremental changes can't match the scale of the ecological crises we must now address. We need bold and systemic transformational processes that go to the root of our relationship to our planet.

Rebuilding the world post-COVID means focusing on changing economic and financial systems, transitioning from short-term unbridled economic and monetary growth to the flourishing of all life for present and future generations. The focus on gross domestic product (GDP) must give way to indicators of true wealth, indicators that factor in ecological wealth and wellbeing. New Zealand, Costa Rica, and Bhutan are already adopting frameworks for measuring human wellbeing and happiness. It has been incredible to see that ideas such as regenerative economies, bioregional self-reliance, economic localization, mutual aid, universal basic income, and debt jubilee—ideas that were considered too radical just a year ago—are becoming mainstream.

The health and wellbeing of people, communities, and the biosphere are interlinked. Our bodies exemplify the patterns of healthy living systems; for example, our cells self-organize in myriad networks to keep us alive and thriving. Nested in an interconnected and interdependent web of life, humans are similarly cells in the body of a living Earth, and thus need to be in service to life. Instead, our species has organized and patterned itself in human communities that are at war with the web of life! If this condition existed in our own bodies, it would be akin to an autoimmune disorder—cells at war with the host organism. Infinite economic growth on a finite living planet is akin to the logic of cancer in a body: cells growing out of control until they kill the host. From the cells in our bodies to the biosphere, aligning with a living Earth worldview is fundamental to our long-term survival as a species.

Nowhere is this logic of life clearer than in the processes that are at play in the Amazon Rainforest, which is a vital organ of our living, breathing Earth. How ironic that a global pandemic causing so much sorrow, pain, and death offers at the same time an opportunity to embrace the living Earth worldview and to align our economic and governance systems in service to life. We can fuse ancient Indigenous wisdom with science-based knowledge to learn how to be good ancestors to future generations of all species; we can see ourselves as relatives to all life forms, a strand in the complex and sacred web of life.[1]

1. I have found certain crucial texts profoundly illuminating; they are must-read books for all who are searching for how to be a productive cell in the body of our living Earth: Capra, *The Web of Life*; Eisenstein, *A More Beautiful World*; Kimmerer, *Braiding*

There is hope and possibility, and a great deal we cannot know yet. But if we seize this moment, we will learn, we will evolve, and we will flourish.

Planetary Health:
Protecting the Amazon as the Heart and Lungs of Gaia

The Amazon is the size of the continental US and contains half of the Earth's remaining tropical rainforests. It is home to thirty percent of all species on Earth and nearly four hundred Indigenous cultures. The Amazon can be called a vital organ of the planet's biosphere—one that sequesters carbon, cools the South American continent, produces oxygen, drives weather systems, causes rain to fall, and regulates the climate regionally and globally. As we better understand the Amazon's role in the Earth's hydrological cycle, we see that it functions as if it were the heart of our planet.

The sheer size and scale of this biome is mind-blowing! The volume of water flowing out from the Amazon River into the Atlantic is greater than the combined flow of the Earth's next six largest rivers (and two of those are Amazon tributaries). More than half of the water cycling through the Amazon Basin is found in the massive atmospheric rivers that bring rain to the entire continent of South America and from there to the rest of the world. These "flying rivers" are fed by the trees, which on average generate up to one thousand liters per day of water vapor per tree. It is estimated that seventy percent of South America's GDP is dependent on Amazon rain.

Rainforests around the world, and in the Amazon Basin in particular, are a primary source of our modern medicines. Indigenous peoples, especially elders, hold vast knowledge of these medicinal plants, whereas less than five percent of these rainforest plant species have been catalogued by science. Consider that an Indigenous elder can recognize and use some five hundred to a thousand types of plants, many with medicinal properties. Among other diseases, the cures for COVID, cancer, and diabetes could well be found in these forests, which contain plants that exist nowhere else on Earth.

The Tipping Point Is Near

Scientists warn that deforestation is driving the entire system towards a tipping point, a point of no return. This tipping point is reached when a forest can no longer generate sufficient rain to sustain itself. At that point it

---

Sweetgrass; Benyus, *Biomimicry*; and Korten, *Change the Story*.

begins to experience a massive dieback process, which over a relatively short period of time will turn the rainforest into a savannah.

According to Carlos Nobre, a member of the Brazilian Academy of Sciences and a lead scientist on the recently formed Amazon Science Panel, "Without immediate action to halt deforestation and start replacing lost trees, half of the entire Amazon rainforest could become savannah within fifteen to thirty years."

The best available modeling predicts that we are perhaps five to fifteen years from the point of no return. Those numbers come from before Brazil President Bolosnaro came to office, however. The devastating fires of 2019, deliberately set to clear the forest for pasture, destroyed approximately 5,400 square miles of the Brazilian Amazon alone, which was a doubling of the prior year's deforestation rates.

When most of the world shut down economic activity during COVID, legal and illegal mining, oil drilling, logging, and clearing of land for agriculture did not stop. On the contrary, taking advantage of the reduction in law enforcement, clandestine activities increased. As if this wasn't bad enough, the Amazon burning season started even earlier in July of 2020. We are perilously close to this point of no return, and in parts of the Amazon, we may already be there.

I believe that Indigenous peoples are the keys to the future of the Amazon Basin. Indigenous peoples have kept the Amazon's forests in good standing for millennia. Indigenous territories across all nine Amazon countries contain some of the best-conserved forests in the Basin. Research shows that average deforestation rates are two to three times lower in Indigenous territories than in adjacent "protected areas."[2]

Indigenous peoples have been successfully fighting the Amazon's destruction for decades. They are engaging in protests and direct action, filing lawsuits and grievances in international tribunals and domestic courts, telling stories, and recruiting global allies. They have blocked dozens of oil and mining projects. Now they are calling on the world to stand with them in shifting the economic paradigm and permanently protecting the most biodiverse region of the Amazon, the Sacred Headwaters. José Gregorio Díaz Mirabal, a COICA Leader (Coordinator of Indigenous Organizations of the Amazon Basin), beautifully describes their vision: "We must save the Amazon and the future of humanity through an economy that respects the life cycles of nature and that recognizes the rights of nature and of Indigenous Peoples. We are seeking an economy that sees life as a whole, and not only for its monetary value."

2. Walker et al., "Forest Carbon in Amazonia."

The Amazon Sacred Headwaters Vision

How will the ravaging economic consequences of COVID show up in the nine countries of the Amazon? Will these countries find themselves going even deeper into the failed extractive economic model and deeper into debt? Or will they integrate the principles of living systems into re-designing their economies, food systems, energy systems, and financial structures?

Over the past three years, I have had the honor of working closely with an alliance of twenty-five Indigenous nations and their trusted NGO allies (Amazon Watch, Pachamama Alliance, and Stand.Earth) as they work to permanently protect the Amazon's headwaters in the Napo and Marañon river basins. The Amazon Sacred Headwaters region starts high in the active volcanoes of the Ecuadorian Andes and spans seventy-four million acres in Ecuador and Peru, making it roughly the size of the state of Oregon. Nearly 400,000 Indigenous people and about 1.5 million non-Indigenous people call the Sacred Headwaters region their home.

The Amazon Sacred Headwaters: Territories for Life Initiative advances ecosystem-based and Indigenous-led governance and stewardship for this vast and beautiful region, working to protect its remaining forests, watersheds, and amazing Andean and Amazonian biodiversity. Similar to the Green New Deal, our Initiative is supporting a holistic, Indigenous-led process to create a "Bioregional Life Plan" for the Sacred Headwaters. Central to the plan is developing innovative ecological-economic frameworks, so that the collective wellbeing of communities within a flourishing Amazon biome can supplant the paradigm of economic growth and industrial development. The Initiative's Declaration describes the process as a "bottom-up participatory process of visioning a future for the region based on the recognition and respect for Indigenous peoples' collective rights, the rights of nature, and the pursuit of collective wellbeing." The Indigenous philosophy of *Buen Vivir* (collective wellbeing), along with cooperation and harmony, are at the heart of this vision for the Amazon's future.

A bioregional planning process is underway to generate, debate, and influence public policies at the national and international level, especially in light of the upcoming elections in both Ecuador and Peru. The emerging vision includes a number of inter-related solution pathways, which include:

- No new oil and mining concessions in Ecuador and Peru
- A globally supported agreement for leaving oil and mineral reserves in the ground
- "Debt for nature" swaps and debt forgiveness agreements with China, the IMF, and other international lenders

- Programs and public policies to reduce deforestation
- Support for alternative livelihoods and regenerative economies, along with sustainable production
- Resources and technical assistance for implementing the life plans of Indigenous nations
- Programs for the protection and restoration of critical ecosystems and ecosystem connectivity
- Support for Indigenous territorial governance
- Remediation plans for areas contaminated by oil and mining
- The empowerment and participation of women in governance and decision-making
- The protection of Indigenous ancestral knowledge
- Sustainable transportation and communication
- Culturally appropriate and ecological education
- The promotion of food sovereignty and local self-reliance
- The shift to renewable energy
- Green intelligent cities and innovative urban planning
- A Sacred Headwaters Fund to support the implementation of Indigenous life plans and livelihoods

By addressing multiple challenges simultaneously and supporting bottom up solutions, we believe that we can help bring about a total system transformation.

We are also exploring innovative concepts in regenerative finance, including Universal Basic Income (or, more exactly, Social Function Income and Job Guarantee Schemes), in order to transition those working in the extractive industries. Possible solutions for financing the transition include alternative systems of exchange and currencies, cooperative lending, positive money systems, linear interest financing, and "eco taxes." Our explorations are guided by the frameworks and principles of commoning, bioregional self-reliance, degrowth, regenerative economies, economic localization, doughnut economics, and ecological debt.

Our vision for the future comes from action in the present. In 2019, Indigenous peoples of the Sacred Headwaters region issued a declaration calling on the governments of Ecuador and Peru to halt the expansion of new fossil fuel projects, mining, and large-scale industrial development in the headwaters regions of the Amazon River, including cattle ranching and road building.[3] The goal is to permanently protect the region as off-limits to resource extraction, which has caused widespread deforestation and biodiversity loss and has led to the destruction of Indigenous populations and resulting human rights abuses.

Over the course of the last ten thousand years, Indigenous peoples have slowly learned the ways of the Amazon rainforest and have cultivated gardens that have enhanced the biodiversity of this vast biome. They know the ways of the forest, can predict weather patterns, sense the movement of animals, and are aware of myriad subtle changes in the forest that outsiders cannot perceive.

Indigenous Life Plans—community-generated visions and holistic plans for Indigenous peoples and their specific territories—play a key role in capturing and reflecting an Indigenous worldview within the Initiative's participatory planning processes. Traditional knowledge must guide the transition. Ongoing conversations with elders, women, and youth are helping them to refine their worldview and to translate their dreams into life-affirming solution pathways. The result is policies that will sustain the Sacred Headwaters for their and humanity's future.

Our Collective Future Depends on the Fate of the Amazon

At the same time that we are sounding alarm bells, it is also important to connect with the Amazon from a deeper, heart-centered place—to see it as a glorious feature of Gaia, full of secrets and teachings, stories and legends, myths and mysteries. The Amazon is perhaps the most magnificent display of nature's intelligence and creativity. The rainforests are alive and teeming with life; they inspire awe and wonder; and they offer to us profound epiphanies about ourselves and our place in the cosmos.

Even if one is not able to be there physically, it is a place we can visit in our psyche in order to hear its call—a vast, unfolding, breathtakingly beautiful, and magical realm that inspires our imagination and creativity and teaches us about the wisdom of nature. The Amazon has an intrinsic value; it has the right to exist and regenerate beyond the value that it provides to us humans. Once we connect our hearts to the heart of Gaia, we will see that we are part of this complex life web and that we can be guided by her in finding our way through COVID and the other crises that we will face.

Someday we may reflect back on this pandemic and see it as a miraculous moment, the turning point toward the birth of an ecological civilization emerging from a Living Earth paradigm.

3. Amazon Sacred Headwaters Initiative, "Indigenous Peoples."

References

Amazon Sacred Headwaters Initiative. "Indigenous Peoples' Declaration for the Amazon Sacred Headwaters." 2020. https://sacredheadwaters.org/declaration.

Benyus, Janine M. *Biomimicry: Innovation Inspired by Nature*. New York: Perennial, 2002.

Capra, Fritjof. *The Web of Life: A New Synthesis of Mind and Matter*. London: Harper-Collins, 1996.

Eisenstein, Charles. *The More Beautiful World Our Hearts Know Is Possible*. Berkeley: North Atlantic, 2013.

Kimmerer, Robin Wall. *Braiding Sweetgrass: Indigenous Wisdom, Scientific Knowledge and the Teachings of Plants*. Minneapolis: Milkweed, 2013.

Korten, David C. *Change the Story, Change the Future: A Living Economy for a Living Earth*. San Francisco: Berrett-Koehler, 2015.

Walker, Wayne, Alessandro Baccini, Stephan Schwartzman, Sandra Ríos, María A. Oliveira-Miranda, Cicero Augusto, Milton Romero Ruiz, Carla Soria Arrasco, Beto Ricardo, Richard Smith, Chris Meyer, Juan Carlos Jintiach, and Edwin Vasquez Campos. "Forest Carbon in Amazonia: The Unrecognized Contribution of Indigenous Territories and Protected Natural Areas." *Carbon Management* 5.5–6 (2014) 479–85.

US

Everything Depends on THIS Depends on Everything **by Nikki McClure**

4

The Architecture of Abundance
A Path to a Democratic Economy

JUSTIN ROSENSTEIN AND THE ONE PROJECT TEAM

Felicia considered the Autumn cycle reports as she walked out into the warm evening. The news was being read and celebrated around the world. Global poverty: down ninety-three percent since 2020. Every climate target hit this year, with three unexpectedly surpassed. Topsoil the healthiest it has been since 1920. Average work hours reduced another eighteen minutes to reach 19.2 per week, with one hundred percent employment of those able and willing to work. And perhaps most hearteningly, average global self-reported life satisfaction up two percent year over year. A bittersweet feeling washed over Felicia as she listened to the quiet of the city amid the human voices, cicadas, and songbirds.

She felt enormously proud of her contribution and relieved that her service on the Regional Council was over. It had been painstaking work, a year-long crash course in ethics, ecology, and sociology. Weeks of debating and ultimately collaborating with people with whom she initially seemed to have little in common. Difficult tradeoffs to weigh, always. But ultimately, they had done their part, as had thousands of other local, regional, and global councils—the mistakes of some compensated for by the wisdom of many others. And the results spoke for

33

themselves: the will of the people made manifest in a world that was better today than it had been a year earlier.

Yet certainty was giving way to an unmapped future. Now proposals were being requested for projects that could contribute to the planting of ten billion trees annually in some of the hardest-to-reach rainforest terrain. Some were just concepts, but others were farther along and closer to being considered. At home, five new playgrounds were being requisitioned, and she could imagine spending some long months near her parents, designing and building them. Or perhaps she would go to Taiwan where her sister was part of a fellowship class training the next generation of AI ethics researchers. There was a Navy position based nearby on the ocean plastics removal team, and she had always wanted to spend time at sea.

Felicia knew what her father would say: "Choices are hard. But having no choice is a lot harder." Her father had had no choice but to sneak across several borders when only sixteen and fleeing from a civil war. He had no choice but to work fourteen hours a day slaughtering chickens in some gruesome perversion of a farm. And he had no choice but to sit in a private prison for a year after immigration services raided the killing floor on which he worked.

"We had four hundred choices of breakfast cereal," he often told her, "but no control of our lives." Her father's world was only twenty-five years in the past. To Felicia it felt almost impossible to imagine.

But her father's world is our world. In 2015, representatives of humanity sat at one table to develop shared goals for a sustainable future, the closest thing we've ever had to a global will of the people. The UN Sustainable Development Goals expressed a soaring vision for what we could achieve by 2030. End poverty. Universal education. Gender equality. Sustainable resource use. Yet everyone involved in the process knew that the global economic system, and the governance systems that support it, had little incentive or mechanism for achieving these goals. The goals would rely on charity and voluntary commitments. It's no wonder then that progress over the first five years has been minimal and funding has been trillions of dollars short, while humanity slides past ever more ecological limits.

We can do much better. I just offered a glimpse of a possible world in which equitable, deeply democratic processes, aided by technology, give power and voice to all people and enable humanity to set and achieve shared goals for the management of our precious planet. It's a world of local autonomy and diverse cultures with diverse priorities that are still capable of coordinating at the global level on issues like climate change and existential threats from technology. It's a world where politics is done without professional politicians, and freedom is achieved without the disastrous side effects of capital markets. It's a world in which we had agreed that poverty, hunger, climate change, and the destruction of nature are evil, recognized that protest and creative resistance had been necessary but insufficient, took matters into our own hands, and changed course.

In the 1980s, Margaret Thatcher convinced much of the globe that "there is no alternative" to the modern market economy. That sounded plausible as the world witnessed the rampant corruption and ultimate collapse of totalitarian communism—the dominant alternative at the time. But Thatcher's statement is obviously false as we look toward the future. The ways we work, consume, communicate, find love, and learn have been entirely revolutionized since Thatcher left office. Is it possible to imagine there is still no alternative to economic and political systems that were birthed in the eighteenth century?

This essay proposes a radical alternative: that we begin to transition from our current world—dominated by amoral markets, backsliding democracies, and neo-authoritarians—to a new system for meeting our needs, wants, and goals that is effective, equitable, and ecological by design. We are advocating nothing less than replacing today's values-blind economy with a values-first economy of which all stakeholders are fully in control. We must transition mindfully, incrementally, in a way that anticipates the unexpected and systematically learns from mistakes. But we must do it, and we must do it soon.

I'm not talking about a retread of failed twentieth-century alternatives like planned-and-command communism or "green" capitalism. Instead, we can use the power of modern information technology, combined with age-old wisdom about cooperative use of common resources (wisdom that market ideology has sought to erase and deny), to create something radically better. Felicia's world is to ours what Wikipedia is to Britannica, what regenerative agriculture is to factory farming, what a pluralist society is to apartheid—something freer, more effective, benefiting everyone, and easy to dismiss as impossible until it takes root.

In order to imagine the systems of the future, we must keep our minds open, and accept that our imaginations can only glimpse the future. Still, we must—with care, humility, and the participation of everyone affected—begin to transition to new systems that optimize for the will and wellbeing of the people and the planet.

~

My whole adult life, I have acted from the deep intuition that collaboration is a key to reducing suffering and improving the human condition. For the first fifteen years of my career, I worked from the assumption that if I could help make collaboration easier and faster, social good would be the emergent result. Starting in my early twenties, I co-led the development of Google products like Drive, Facebook features like the Like button, and then Asana, which helps teams of all kinds work better together.

Asana and Drive are tremendously useful at helping teams achieve their goals more quickly. (My feelings on social media's evolution are more complicated.) And many organizations use Asana to do great things in the world.

Yet, over the years, especially as I came to increasingly question the direction humanity was taking, I asked: Where are team goals coming from in the first place? On examination, the primary forces determining organizations' objectives appeared to be modern systems of economics, governance, and sensemaking that—while extraordinarily capable at delivering certain kinds of goods—are insufficient for the problems we now face, and only perpetuate colonial-era class, race, and power structures, degrade the natural world, and exploit and neglect billions of people.

This has been obvious to many for a long time. But it wasn't until I started to question the story of perpetual human progress I had learned in school that I began to observe society differently. Each major human system—like food, water, justice, education, healthcare, technology, media, business, community, international relations, defense—seemed broken in a unique way. But our inability to fix them—and, in most cases, the problems themselves—seemed to have a common cause: each system is in the grips of the market-state power complex. How could a surplus of food coexist with a billion hungry people? Why does American industry spend fifty times more money on treating diseases than curing them? It felt heartbreaking and nightmarish that the planet on which I was born was enchained by a giant prisoner's dilemma: The world could work for everyone, and work better for everyone, if we collaborated toward mutually beneficial outcomes, as one team. And yet we weren't.

So a year ago, I started One Project, a non-profit venture to explore what would be required to realign humanity's resources and efforts with our common goals and values. For the past year, the One Project team has been reading widely, while meeting with and learning from a diverse set of critics, visionaries, and leaders—inquiring into the profound philosophical and technical questions necessary to approach generationally difficult problems about how we live life on Earth.

The stories I'm about to tell of a better future, and the features of a system that make it possible, are nascent fruits of this inquiry. We're sharing them now in the hopes that you'll critique and improve our thinking dramatically—and maybe even collaborate with us.

∾

2020 had taught Felicia and indeed all of us that pandemics weren't some far-off what-if. They could arrive at any moment and entirely upend an unprepared world, destroying bodies, but also ripping social fabric, destabilizing governments, driving into the ground businesses that took generations to build. So, when we transitioned away from a values-blind market-based economy to a goals-based one, Felicia, her regional Citizens Council, and indeed Terrans everywhere identified pandemic preparedness as a goal needing significant attention and investment.

For starters, Terrans built a public global knowledge base, a wiki populated with all we had learned from COVID-19. Over the years, epidemiologists, geneticists, sociologists, survivors, frontline workers—and even those who questioned whether there had been a dangerous outbreak at all—added to it. A compendium of complex, conflicting information. The technology didn't eliminate differences of opinion, but it facilitated productive dialog, managed debate, and built consensus; and like its primitive ancestor Wikipedia, it gave an imperfect but still fairly sharp picture of our global collective intelligence. This process of making sense of the world made politicized briefings and cable news pundits look like the ancient relics they had now become.

When it came to the Pandemic Preparedness Global Council, few knew exactly which experts would be most capable and trustworthy to coordinate a future response. But everyone knew what a disaster 2020 had been. Because of widespread mistrust of government, vast swaths of the population had believed that

officials in charge were incompetent puppets or dark conspira-
tors. Social media companies, in turn, made billions spreading
lies and innuendo about these well-meaning experts. But this
time, the experts were of the people's choosing. Citizen Councils
like Felicia's and the people they served each had placed trust in
the most informed, high-integrity friends, acquaintances, and
leaders in their own circles, who in turn had given their trust to
yet more informed experts through a well-honed process known
as "liquid democracy." As trust flowed organically to those with
the most specialized knowledge and integrity, a global team was
assembled that held real legitimacy in the people's eyes.

The Council, in turn, drew from our shared intelligence to
allocate mass resources around the globe to build PPE stock-
piles. Better mask designs were pioneered out of Chennai. The
instructions for making them became instantly available, and
distributed manufacturing projects launched in every bioregion
to take advantage of the innovation. Funding was allocated to
contact tracing infrastructure that was customizable to meet
each community's standards for privacy. Smart thermometers
that tracked unusual illness activity in neighborhoods were
distributed widely. At the local level, Felicia and her fellow
citizens didn't each keep their eyes on every program, of course.
Everyone was busy living their lives. But the Council, like every
council, reported to *us the people,* all people, and every move
they made was transparent, open-source, and open to feedback.

So when COVID-32 exploded, the world was ready. The
Council proposed an immediate two-week worldwide pause.
In 2020, the global economy was like a shark—it would die if
it stopped moving forward. Back then, prioritizing lives over
economic activity meant material insecurity for the most vulnerable workers, destruction of massive fictional value in capital
markets, widened inequality, and broken dreams everywhere.
Was it any wonder that countries struggled to attain compliance
with lockdowns? This time, Felicia's region, and regions around
the world, were ready with community-based support systems
that reduced economic fear, and there were no longer incentives
for powerful business interests to push for hasty re-openings.
Throttling down would mean other important goals would be
deferred, but without the need for anyone to lose their livelihood or their home. We were thinking long-term. So when the
Council recommended a pause, most people paused, while resources were redirected to protect and generously compensate
the essential workers who couldn't.

Over the next six months, as the virus simmered and occasionally flared, there was a flurry of activity. In Felicia's city, thousands of us in good health transitioned our work lives—getting groceries for our neighbors or teaching kids whose parents were essential workers. Those unable to leave their homes signed up to provide emotional-support calls to the most vulnerable and alone. In the evenings, many served on the peer-to-peer response service, answering questions, aided by the Pandemic Wiki, about everything from safe socializing to battling feelings of depression. Others poured over data on the effectiveness of interventions, played prediction markets to anticipate the next flare-up, or home-schooled quaranteams of children. Skills were matched quickly to needs, as we banded together as a community. Not long ago, this would have all been called "volunteer work." But in a goals-based economy, those performing needed work were compensated well.

As we protected our communities, teams around the world divided up the work of cooperatively developing treatment protocols. Those that showed the most promise applied to the Treatment Council for more resources. Dead ends were abandoned. We developed powerful medicines, quickly. But finding an effective treatment turned out not to be our most difficult challenge. The disease was highly contagious and virulent—everyone wanted to be protected. While the new medicines belonged to the people, initially there wasn't enough to go around. Deciding who got treatment first was an ethical dilemma for which councils of scientists weren't suited. Instead, representative Citizen Assemblies, everyday people chosen by lottery, convened in each region. With consultation and education from broadly trusted scientists, ethicists, and network theorists, each assembly determined a treatment distribution strategy that was culturally appropriate, legitimately democratic, and just.

COVID-32 was indeed the super virus for which COVID-19 had been a dress rehearsal. But globally, there was far less death and far fewer shattered lives than the last time around. We mourned. We implemented processes to learn from our mistakes. But we also celebrated. Some credited better data availability or easier partnerships or better aligned incentives for our success. Others said it was the reduction of weaponized misinformation, and more democratic processes. In fact, it was all of these, but each was a different flower nurtured by the same root—a truly democratic economy based not on an insatiable fiction called profit, but instead on what was good for people and the planet.

~

Imagining a future like this helps us realize that our biggest challenges can be addressed. Consider the current destruction of the biosphere on which all life depends. As long as our economic system allows for unaccounted ecological externalities, without a reliable path to a systemically supported turnaround, we are at exponentially increasing risk of destroying our only viable biosphere—and very soon. Today, the market doesn't demand that anyone pay for most pollution, destruction of ecosystem services, degrading of resources for future generations, or catastrophic climate change; yet all of these are the result of economic activity. Attempts to hold polluters accountable have been counter to the dominant religion of economic growth, and so they have never had significant impacts. A goals-based economy would let the people, in consultation with experts, build accounting models that count all the good *and* all the harm created by economic activity (and track them through supply chains via sensor-enabled logistics tech). Producers would compete to minimize their negative impacts on other important goals or even design their activity to have positive outcomes. Well-resourced start-ups would collaborate and compete to develop the best holistic solutions for carbon drawdown. In such an economy, a stabilization of the biosphere would be part of a new definition of economic growth, growth of the things that make us and our planet healthy.

We can also imagine a far more equitable future, in which the distribution of goods works very differently. Financial markets are an imaginary game, and money is their scoring system. In capitalist mythology, a person's score reflects their grit, talent, and frugality. In reality, scores are rigged. Since the thirteenth century, Europeans have been "enclosing" common lands (and people) around the world with fences and violence, calling them "property" and "capital." Unsurprisingly, to this day, their descendants have more points in the game.

We can transition to a new game, where points can be gained only through genuine contribution to (or need from) society, rather than through renting, loaning, extracting, exploiting, commodifying, abstracting, or "middle-manning." We can design the mathematics of the game to be fair: the people who contribute the most, get the most, within democratically-established bounds of equality, compassion, and reparation.

In the old game, unemployment and poverty are natural, even necessary. In the new game, there's always work (and skills training) to go around, with fair compensation that respects everyone as peers, and software that assists in matching you to the role in the one human project that's the best fit for your capacities and curiosities. In this world, immigration means more contributors, people who lighten others' loads and measurably expand the size of the pie and everyone's piece of it. Automation is not a threat, but a tool to reduce tedious labor. We can all enjoy its fruits and spend more time with the people, places, and things we love.

This isn't state socialism. Here, property is not owned and distributed by governments. It is understood rightly as the wealth of the people and the planet, allocated and apportioned by distributed democratic processes and the formulas upon which they converge.

We live in a bizarre and unjustifiable world in which the color of one's skin likely determines the length and quality of one's days. Today's economy is unabashedly plutocratic and implicitly racist, sexist, ableist, ageist, and colonialist; tomorrow's economy could be deliberately reparatory and democratic. To avoid recreating the oppressions that have been part and parcel of liberal democracy, such as the tyranny of the majority and the silencing of the oppressed, a new system can empower demographically representative groups of ordinary citizens to meet, deliberate and make decisions, guided by constitutional frameworks that specifically resist oppression. Groups tend to blame or victimize others in the absence of positive, personal experiences with one another. Bringing diverse councils of community members together, online and face-to-face, in inclusive, facilitated processes has been shown to generate mutual understanding and creative problem-solving, avoiding the typical win-lose dynamics of our current democracies. In such environments, difference is not something to be resolved or overcome. Rather it becomes the very source of higher levels of group intelligence and a more complete model of reality.

In the new game, public institutions report to the people they most affect, not to distant politicians and their appointees. When the people control the funding, rules, and leadership of the justice system, we can begin to heal from collective traumas like mass incarceration and the war on drugs. We can take back control of the designs and resources of our neighborhoods. (And, in a fair society where everyone has enough, there will be vastly less need for policing.)

When my team and I share this vision of a humanity capable of taking on its greatest challenges through deep collaboration, we usually encounter a few objections.

The first is that mass coordination is too complex. The myth of today's system is that it's simple, even natural. People have needs. Other people meet those needs. Everyone's free, and the Invisible Hand makes the world go round. In reality, if my infant daughter asks me about our current economic system when she gets a bit older, it will be mind-numbingly complicated to explain. She could truly understand its design only after adopting a constellation of man-made social constructs: private ownership, "money" as abstract wealth, compensation based on extraction rather than creation of net value, compound interest as a reward for debt-creation, taxation, the Leviathan state as property enforcer, the "fractional reserve banking" Ponzi scheme. And I would fail entirely at explaining why people tolerate what its structure requires and generates: artificial scarcity, artificial demand (today's global-scale brainwashing by corporations), unnatural poverty, rampant inequality (a mathematical inevitability of free markets), and utterly unsustainable consumption of resources.

The ideas we've shared here leave us with many questions. But we believe there is a simple elegance to a political economy in which people jointly set goals for the good of their local and global communities, and then allocate resources and efforts to reach those goals. This is not a new or exotic concept. The UN Sustainable Development Goals are an implicit global call for goals-based economic management. Indigenous and non-industrial societies have managed common resources in this way for 99.9 percent of human history. Even today, 2.5 billion people depend on forests, fisheries, farmland, irrigation water, and hunting grounds that are managed as commons. Open source software communities have used commons-based thinking to build the backbone of the internet. And broadly, in the history of the evolution of life, cooperation out-competes competition in the long run. Any new system will be too complex to sketch on a napkin, but we think the current system can be vastly streamlined to serve what we, the People, most desire.

The second objection is that such a system sounds like a techno-utopian fantasy that, like all such fantasies, will ultimately prove dystopian. We too are worried about technology's impact on society. It is our current political-economic system, however, that makes technology so dangerous. In today's world, technology is created by an elite few for the ultimate benefit of an elite few. So-called "free" social media services employ self-improving algorithms to hold our attention, manipulate our emotions, and change our behavior for the profit of advertisers and political actors. Their trillion-dollar market caps attest to their success. Artificial intelligence is advancing by leaps and bounds to put more power in the hands of corporations who are locked in an arms race of competition for people's mindshare, money, and loyalty. The brave few computer scientists and ethicists working on containing these technologies rely on non-profit pittances in their efforts to protect the future of humanity.

We propose putting the governance and ownership of technology in the hands of all people. Algorithms and AI can become our allies in translating our human values and aspirations into executable work, aiding our collective intelligence and allowing us to do more with less pressure on our planet. This vision is far from a technocracy. It is democracy realized to a far higher degree than ever before—direct participation by all people, assisted by technology.

As radical as all this sounds, the seeds of this future are being planted today. Imagine an accessible, user-friendly website and mobile app, built as a commons-based, open-source platform cooperative, that allows any community to govern itself through collective intelligence, productive deliberation, and intelligent resource distribution. It's informed by the visionary work of existing practitioners of liquid democracy, citizen assemblies, trust graphs, participatory budgeting, open-source designs for local manufacturing, and circular economies, to name just a few. It is decentralized and secure thanks to protocols like Holochain. If such technology feels far off, consider that collectives across the world are already building prototypes of platforms that provide these services and more. In 2020, Taiwan gracefully managed its COVID outbreak through digital democracy tools that built trust and leveraged citizen participation in ways similar to the stories we told above. But existing technologies only hint at what's possible.

Once we're up and running, any community could set shared goals to work toward, agree upon equitable incentive systems, distribute work fairly, scale collective sensemaking, and build bonds of trust. All of these activities are social and human-centered: they're about people, connecting with people. Their effectiveness can be vastly amplified through forms of real-time communication, information processing, and data sensing that were nearly inconceivable just thirty years ago but are commonplace today. It is heartening to see that, just as quickly as our current systems seem to be failing, our capabilities to replace them are growing.

Such a platform, of course, won't transform the global economy overnight, but it could be adopted in the short-term by communities eager to embrace something new. These may be communities of place, like villages, towns, cities—including some of the hundreds of new cities that will soon be needed by climate refugees. Or they may be geographically far-flung, like social movements. The more people who feel let down by the broken promises of legacy systems, the more will be ready to migrate to something new. If, together, we create systems that can outcompete the old in delivering true prosperity, empowerment, and life satisfaction, we can scale up. We can help build movements of people who have tasted true democracy and who are calling for its use to expand. Such movements can win elections that bring the will of the people to power and put state resources under truly democratic control. Once communities create a new possible, they will never again be convinced that "there is no alternative."

Humanity has no roadmap from today to what futurist Kevin Kelly calls Protopia, a world that gets progressively better every year—healthier, just, safe, creative, and fun. What we do know is that calls for structural change and the terminal flaws in the status quo are now impossible to ignore. Some will choose to double down on the thinking of the past. Some will seek to patch it up.

Still others will see the chance to co-create a new reality based on values like justice, fairness, empathy, solidarity, non-violence, compassion, interdependence, and love for the universe. We intend to work with these values-based allies to help write the next chapter, while listening for the future that wants to be born.

∽

This essay was a collective labor of love. We would love to hear your candid reactions, feedback, concerns, criticisms, ideas, and visions; show you the online (eventually, wiki) version of this essay with case studies and references; and let you know when we release new essays and products—all of which can happen at http://oneproject.org/thenewpossible.

Thank you to the many people who generously gave their time to providing comments and suggestions. This essay is theirs too, and our future work will integrate even more perspectives.

5

Ubuntu

The Dream of a Planetary Community

Mamphela Ramphele

In the 1970s I participated in a student uprising with many others, in which we confronted our own complicity in our oppression. We reclaimed our humanity and our culture, including our African names and languages. Previously, we had accepted the standards set by our oppressors—they decided what was of value and what was not. As activists, we rejected being called "non-Whites" and "non-Europeans" and identified instead as black and proud.

The 1976 Soweto Uprising by black high school students was the spark that led to a mass democratic movement, which ultimately liberated us from apartheid in 1994. The dawn of democracy offered both black and white South Africans an opportunity to heal the wounds of racism manifest in inferiority and superiority complexes. While the offer has yet to be accepted on a wider scale, the embers of that spark still smolder.

Moments of existential crisis bear within them the ability to dream and imagine new possibilities. They contain the opportunity to see beyond the self-imposed bounds of what is possible and embrace a new horizon. COVID-19 has revealed a space where human community can go beyond our comfort zones and reduce the risks we face together. The impact of behavioral change on the scale we have seen the last few months is shocking to many, but this shift reflects the untapped capacity of human beings to change in response to an existential threat.

The extensive behavioral changes have been about more than personal survival, such as wearing protective masks and gloves. They have unleashed a reservoir of compassion and reaching out to those around us in distress, those who are in need of food, care, and protection. The "we are in this together" sentiment has been widely shared, especially in the early days and weeks of the pandemic. We showed up with the best face of humanity: generosity and solidarity.

I suggest that as we draw from the well of generosity and solidarity within us, we accept an invitation to reclaim the essence of our "humanness." This essence lies deep in the souls of each living human being. Kofi Opoku, an African scholar and elder descendant of the Akan people of Ghana, expresses this more eloquently: "The concept of human beingness, or the essence of being human, termed *Ubuntu* in the Bantu languages of Africa, is central to African cultures and religious traditions. It is the capacity in African culture to express compassion, reciprocity, dignity, harmony, and humanity in the interests of building and maintaining community."

Mutombo Nkulu-N'Sengha, another African scholar of the Democratic Republic of the Congo, elaborates further to show how this concept finds resonance in the wisdom of other cultures across the globe:

> *Bumuntu* is the African vision of a refined gentle person, a holy person, a saint, a *shun-tzu*, a person of *ado*, a person of Buddha nature, an embodiment of Brahman, a genuine human being. The man or woman of *Bumuntu* is characterized by self-respect and respect for other human beings. Moreover, he/she respects all life in the universe. He/she sees his/her dignity as inscribed in a triple relationship: with the transcendent beings (God, ancestors, spirits), with all other human beings, and with the natural world (flora and fauna). *Bumuntu* is the embodiment of all virtues, especially the virtues of hospitality and solidarity.[1]

These concepts of what it means to be human are not confined to African cultural and religious traditions. They are shared across the globe. Indigenous peoples in India, the East Asian region, the Pacific Islands, and the Americas have an understanding of much of the same core essence in their worldviews. This should not be surprising given the common origins of humanity and the shared heritage of the mother continent of Africa. At the same time, African religious wisdom does not claim to be the only and ultimate truth but remains open to other impulses. As the Shona proverb

---

1. Nkulu-N'Sengha, "Hope of Liberation," 222–23.

goes: *Truth is like a baobab tree, one person's arms cannot embrace it.* This openness is a sign of humility that encourages conversations across the world we inhabit as a human community.

The power of the concept of *Ubuntu/Bumuntu/Iwa* (Yoruba) and *Suban* (Akan) appears in the way it is seamlessly integrated into a way of life. Conversations and social engagements across generations are opportunities to shape the personal characters of children and adults; they reflect the values of *Ubuntu* as a way of life. I was blessed to grow up in a large extended family in rural Limpopo Province of South Africa. It took the whole village to raise us as children and to understand what it means to be a person reflecting these values.

We were compelled to consciously shape our character and measure our behaviors against the gold standard of *human beingness*. Good behavior was reinforced by public pronouncements by the elders: "this kind of behavior shows that one is a real person." Bad behaviors toward others and other forms of life—animals or plants—was sanctioned through public reprimand: "a real person does not do this sort of thing." Adults reinforced good, responsible behaviors by modeling such actions and by nurturing an acute awareness of the expectations of our families and communities about what behavior met the gold standard of *Ubuntu*.

The core of the African concept of *Ubuntu* is that one cannot be a complete human being without the reciprocal affirmation of other human beings—*umntu ngumtu ngabantu*. The Akan of Ghana would say: *onipa na oma onipa ye onipa*: (it is a human being who makes another person a human being).

Beliefs and Value Systems
Inspired by *Ubuntu/Iwa/Suban* Distilled

The often quoted "I am because you are" is pregnant with the profound meaning of the generative essence of being human:

- We are endowed with a divine spark that never dies because it is connected to the *source* of all life. "The dead are never dead." Our ancestors are forever part of us.

- Humans are social beings. We are wired to be with others, to nurture and shape who we are, and to make sense of our world with one another.

- Our personalities are shaped by what those close to us affirm or sanction. We are whole and endowed with the potential for right and wrong. We are choice-making beings who are socialized to seek what is right.

- We have the capacity to express compassion, reciprocity, dignity, self-respect, and respect for others. These qualities are inherent in us and define our *human beingness*.

- Harmonious relations within family, community, and society are expressions of character beyond an individual person. The individual is both shaped by and shapes relationships with others to sustain life beyond the self. This is the expansive aspect of the "I am because you are."

- Our human connectedness goes beyond present relationships. We are inextricably related to our ancestors, who continue to live in present generations as guiding spirits. We stand as bridges to future generations who may still be carried as seeds in our bodies and who arrive as children born into this unending web of intergenerational connectedness.

- The interconnectedness of all life makes our being possible and demands that we contribute to its sustainability. African people are totemic in that they affirm their connections to nature by identifying with an animal and/or a plant that best reflects our clan identities. For example, the Rampheles are Bakwena; *kwena* (crocodile) is our totem animal. We also have the willow tree as our totem tree. Our clan's history is associated with water and crossing rivers with the help of our totem animal and tree.

- We have an ecological commitment to conserve and enrich. Our capacity to empathize is the core of our being and essential for the sustenance of life itself. The intimate totemic relationships with plants and animals reinforce our reverence for nature, of which we are a part.

What Contributions Can the *Ubuntu* Value System Make to the Dream of a Planetary Community?

The multiple planetary emergencies facing us today are reflections of our deviant behaviors as a human community. We have strayed from *Ubuntu/ Suban/Iwa*. We have fallen short of the expectations of the *Ubuntu* values in our management and use of nature's resources—hence the planetary emergencies that are upon us. Our conflict-ridden social relationships and exploitative approaches diverge from the values that embody self-respect and respect for all life in the universe. We have severed the inextricable links and interdependence between ourselves, others in the human community, and the whole of nature.

The question we face now is: Having glimpsed the greatness of our inner capability to return to the source of our being, are we ready to reimagine our relationships as humans with *all* life on Mother Earth? Could we dare to dream ourselves into a Planetary Community that can live in harmony as interconnected and interdependent beings?

We must begin by confronting the glaring contradiction presented by an African continent that does not model the beauty of this amazing concept of *human beingness* in its political, social, and economic relationships. Given the richness of its heritage, why is there a rupture between *Ubuntu/ Iwa/Suban* values and the current state in many African societies?

Africa's story is an evolution of traumatic ruptures: slavery, colonial conquest, and post-colonial exploitative governance and socio-economic systems. Each one of these manifold ruptures used dehumanization to break the will of its victims. Current neuroscience tells us that of all the forms of intergenerational traumas, humiliation has the most devastating impact. Africa's humiliation by racist conquerors and their missionary handmaidens devalued its culture and heritage. These efforts obscured our inheritance of God-given black beauty. Racism persists to date and justifies color-coded inequity across the globe, adding salt to these festering wounds.

Martha Cabrera is a psychologist who worked in her native Nicaragua in the aftermath of Hurricane Mitch in the 1990s. She confronted the inability of the Nicaraguan people to rebuild their lives after the protracted war of independence against the US-backed brutal regime. After intensive conversations with people on the ground, she came to the following conclusion: "Trauma and pain afflict not only individuals. When they become widespread and ongoing, they affect entire communities and even the country as a whole . . . The implications are serious for people's health, the resilience of the country's social fabric, the success of development schemes, and the hope of future generations."

As a human community, we have seriously underestimated the impact of multiple intergenerational traumas on people—and especially on those whose worldview rests so heavily on the assumption that our humanity is authenticated in its affirmation by other people. The systemic denial of African humanity that was structured into the process of colonial conquest, combined with slavery across the continent, has devastated the very foundations of African traditional belief systems. It has severed the links Africans believe are inextricable between people simply because they are part of our shared humanity.

To add insult to injury, colonial conquerors denied the validity of the African religious belief systems that affirm the presence of the divine in all living beings. The replacement of these values and this way of life with a god who demanded changes to our culture, including assuming foreign so-called Christian names, was the ultimate cultural genocide. Africans underwent a radical change of culture and language; they not only lost their land and their dignity, but also the anchors of their belief systems. Opoku reminds us of the wisdom of our ancestors: "A person who is dressed in other people's clothes is naked, and a person who is fed on other people's food is always hungry."

Africa, like many regions in the world where Christianity holds sway, has allowed itself to be dressed in other people's clothes. Africans embraced others' languages as the official mediums of communication, even changing Indigenous names to accommodate those acceptable to the gods of the conquerors. Africans eat foreign foods that will never satisfy their hunger. Africa has yet to express itself fully as a contributing member of the community of nations. It has yet to bring its rich heritage of *Ubuntu/Suban* into the circle of ideas of the world community and thereby give shape to global culture.

My own country of South Africa remains a deeply wounded society. Its frayed social fabric is reflected in high levels of interpersonal, domestic abuse of women and children and in abuses of power and public resources at the national governance level. Our human rights-based Constitution with *Ubuntu* as the core principle has yet to translate into our post-1994 society's way of life. We are missing the political will to mobilize society. We must commit public and private sector investments toward healing narratives and practices of our centuries-old wounds. Uprooting abuses of power and public resources by successive post-apartheid governments would be a significant indicator of the healing process.

The same applies to most of post-colonial Africa. Those working with Aboriginal people around the globe have documented similar manifestations of multiple intergenerational traumas. Indigenous peoples across our world continue to live as marginalized minorities in their own native countries: North and South America, Australia, Canada, etc. We need to work together to support the healing of those wounded as part of the process of regenerating our human community and our planet.

What Is to Be Done?

The first step in any healing process is the acknowledgement of the wound-edness that is plaguing us. As I mentioned at the beginning, this was demonstrated in the 1970s during the student uprising. This psychological liberation of the self from the prison of inferiority complexes imposed by other people, who consider themselves superior, is an essential step to freedom of the spirit. As free spirits, we became unstoppable as liberation fighters. The 1976 Soweto Uprising contributed to ending apartheid. These movements offered both black and white South Africans an opportunity for healing the wounds of racism in our country.

The second step requires sustaining self-liberation and transforming the value system of the entire society toward healing. Transforming value systems of societies is hard work. It requires investments in intergenerational conversations and narratives that enable people to externalize their fears and anxieties and to rekindle the life spark essential to living a full life. The logic driving the existing socio-economic system needs to be identified and changed to align with what matters within the value system of *Ubuntu*. In this worldview, human life and healthy environments that sustain life take center stage. For *Ubuntu*, the wellbeing of people and the biosphere is the measure of success in all social endeavors. Our education and healthcare systems, our human settlements and energy systems, our livelihoods and resource systems, should also reflect *Ubuntu* values. The COVID moment has heightened the urgency of this second step not just for South Africa, not just for the whole African continent, but for the entire world.

The third step is the call for Africa to pay more attention to its ancient history as part of the healing of its wounds. The words of John Henrik Clarke, an African American historian, come to mind:

> History, I have often said, is a clock that people use to tell their political time of day. It is also a compass that people use to find themselves on the map of human geography. History tells a people where they have been and what they have been. It also tells a people where they are and what they are. Most importantly, history tells a people where they still must go and what they still must be.[2]

The misteaching of history, not only in Africa but globally, can rob our children of the opportunity of finding themselves on the map of human geography as proud global citizens. African leaders need to invest in transforming the teaching of history to enable African children to take pride in

2. Clarke, "Why Africana History."

their ancient origins and rich heritage. A transformed African history curriculum opens our eyes to the prowess of African ancestors as reflected (for example) in ancient Egyptian civilization. It reclaims Egypt geographically and emotionally to the African continent and to its Nubian origins, freeing it from its imposed identity as a Middle Eastern country. Reconnecting ancient Egyptian history is an essential part of the healing process for the African peoples.

Cheikh Anta Diop, a Senegalese polymath, devoted his life to documenting ancient African history despite huge resistance from some traditional Egyptologists. He concluded that "Ancient Egypt was a Negro (Black) civilization. The history of Black Africa will remain suspended in air and cannot be written correctly until African historians dare to connect it with the history of Egypt." It is noteworthy, though not surprising, that ancient Egyptian culture centered on *Maat*, whose seven principles mirrored *Ubuntu*: truth, justice, harmony, balance, order, reciprocity, and propriety.

Conclusion

Our COVID moment offers us an opportunity to rediscover who we are as a human community. The slowing down of our frenetic, consumption-driven lifestyles has enabled us to look deep into ourselves as a human race. It is a moment that may well go down in history as a turning point for us to come to grips with who we really are as human beings in the larger scheme of our world. This is a necessary process for reclaiming our *human beingness*.

The resurgence of racism across the globe is an indictment against us as a global community. The science we have accumulated and practice affirms the *Ubuntu* notion that there is only one race: the human race. Racism is perpetuated by our willful ignorance and is used to justify a system of color-coded marginalization of those we "other" to promote avarice and inequity. COVID as an equal opportunity invader has challenged us into understanding that we are part of a single human community that has the capacity to work together for the common good of all people and our planet.

Ubuntu, the recognition that "I am because you are," is the horizon of possibility before us. We need to continue this journey and travel deep into our beings, where we are connected to one another and to those who have gone before. We need to pay due reverence to the spark of life inside each of us, and continually raise our consciousness to the sacred light of life within us. This sacred light calls us to reflect deeply on our collective responsibility to shape a future worthy of those yet to be born.

References

Clarke, John Henrik. "Why Africana History?" Department of Africana & Puerto Rican/Latino Studies, Hunter College. Accessed August 14, 2020. http://www.hunter.cuny.edu/afprl/clarke/why-africana-history-by-dr.-john-henrik-clarke.

Nkulu-N'Sengha, Mutombo. "The Hope of Liberation in World Religions." In *African Traditional Religions*, edited by Miguel De la Torre, 217–238. Waco, TX: Baylor University Press, 2008.

6

The Future Belongs to Us, Here's What We're Going to Need

Reflections of a Teenage Activist

Anisa Nanavati

MY HISTORY TEACHERS HAVE told me that our generation is living in a time that will forever be remembered in the history books, and yet, like many others, I am honestly tired of it. In my opinion, everything started to go downhill after the election of one of the worst bigots that our world has known. 2019 had passed after what seemed like an eon; 2020 was going to be a year of opportunity. Blissfully ignorant of what was coming our way, we rang in the new year and wrote our resolutions. This was going to be the year where we finally made a better world.

Our optimistic expectations were interrupted only two days into 2020. We were quickly snapped back into the real world when a United States airstrike killed one of Iran's top generals. As if the looming threat of a world-power face-off wasn't enough, the universe decided to throw in a devastating pandemic that would change the lives of all who lived through it.

Then Came COVID

COVID-19 was initially dismissed as a minor threat. Media outlets were slow to cover the outbreak of COVID-19, and much of what they spread was misinformation. The first that I heard of the pandemic was in an unlikely place, TikTok, a social media site. If you aren't too familiar with the app, it's a popular platform where teens create easy to learn and addictive dances and make incredibly obscure jokes. App creators started to make jokes about the virus in mid-January and urged their viewers to wash their hands. These jokes were all in fun; COVID-19 seemed far off, distant, and unable to touch any of us. No one expected what would come next.

In an unlikely role reversal, social media platforms associated with spreading misinformation and "fake news" actually gave us a real insight into what was happening. Social media platforms like TikTok were used by global organizations such as WHO and the United Nations to inform younger viewers about COVID-19. As the rates of infection grew exponentially in Wuhan, and people started to understand the severity of the disease, the jokes stopped.

We saw videos of Wuhan's doctors and healthcare workers weeping as they begged for worldwide organizations to help them. The streets of a once bustling city were empty and ghostlike. The most striking video I saw was of an elderly woman who climbed down her eight-story building to escape the lockdown. The videos were dismissed by many world powers. They all downplayed the extreme severity of the virus.

I found it hard to believe that a virus that had the power to completely interrupt our ways of life was just like the common flu. These harrowing experiences from people from around the world were not reported on by the regular media, so I again turned to my phone to gain first-hand insight into the impact of the virus. As well as being an excellent timewaster, my phone helps me to stay connected to the world.

Social media has always been a community and place for discussion, but it wasn't always a forum to debate politics. Though social media is dismissed by many as mindless, it has transformed with the growth of the feminist movement and the Marjory Stoneman Douglas High School shooting. Social media has become a place where we can have difficult but necessary conversations as well as take a break from the downpour of graphic images and gut-wrenching stories reported by mass media. The world is a dark place, and our access to information has caused my generation to have some of the highest rates of depression and anxiety. But social media can also be a refuge to turn an incredibly painful experience into something a little more lighthearted and to help ease the pain. The outrage of students affected by gun violence sparked March for Our Lives, one of the largest youth-led activist movements. March for Our Lives used social media to promote events, find organizers, and unite students from all over the United States. Social media platforms are now bustling hubs for learning, debating, and organizing for everything from gun violence to COVID-19 to climate change.

Youth-led organizations used social media to circulate alarming information about COVID-19 that was not common knowledge. The hundreds of posts on my feeds from these organizations, organizers, healthcare workers, and ordinary civilians showed that the outbreak in Wuhan was devastating. But if someone had told me that COVID-19 would continue to ravage through the world, reach the United States, and completely change my life, I would not have believed them.

As the weeks passed, my news feeds became flooded with images of countless body bags and mass suffering all over the world. Healthcare workers became our new heroes. China, Italy, and Spain were devastated by the virus, and it showed no signs of stopping. I could not believe this was our new reality. Seemingly isolated cases of the virus had started to pop up in the United States, but we were told that we had nothing to worry about. Misinformation and false assurances were spread by the very government officials whom we were told to trust.

No one had seen a worldwide pandemic of this scale since the 1918 Spanish flu pandemic, which was heavily downplayed while the United States was fighting in World War I. Scientists at that time had limited knowledge, which made it extremely difficult to combat the flu. There was no precedent for a pandemic then, and there isn't a precedent now.

By early February 2020, the United States government declared a public health crisis. Major outbreaks in Washington and New York led to the first closures of schools. Two days after COVID-19 was declared a worldwide pandemic, the rest of the nation followed, and all schools in the United States closed down on March 13th, which was eerily creepy because it was Friday the 13th.

I remember the air of uncertainty and panic as we left our school in Florida, unaware that day would be our last in the school year. As we all sat in the gym bleachers, our principal gave us a speech on the importance of our civic duty to quarantine ourselves. We asked our teachers every class period about how distance learning would work, digging for more information. It was frustrating that they knew just as little as we did, which was pretty much nothing. We said uneasy goodbyes to all our friends, especially the seniors, and tried to unpack the day. It became our civic duty to give up the one thing that humans depend on, social interaction. It was a sacrifice, but the months in quarantine forced us to pause our lives and adapt to our new reality.

What was most difficult was the transition to online school. My teachers were incredible during this transition and did everything they could to ensure that we were learning. I know this was not the case for every student. The inequalities within our educational system made it incredibly difficult for students without computers or strong internet connections, or those who had busy households, to transition to online school. Many of my friends had limited or zero contact with their teachers, yet they were expected to complete their school work and prepare for the dreaded Advanced Placement tests. The College Board's decision to administer Advanced Placement (AP) tests was a grave mistake. The new AP test consisted of half of the year's material and one or two questions, making it harder to study for and completely unlike the test we'd prepared for before schools closed. Many exceptional students were robbed of the scores that they deserved. COVID-19 gave a clear indication of the changes that we need in our educational system.

Unfortunately, many people were unable to sacrifice even small things for the greater good. Almost every person has been affected or knows someone who was affected by COVID-19. Yet many people decided that their game of golf, their haircut, and their nail appointments were more important than the lives of their fellow citizens. Anti-quarantine and mask protests broke out throughout the nation. These protests blocked emergency vehicles from reaching hospitals and forced the heroes of this pandemic, our healthcare workers, to intervene. Massive parties during spring break, Memorial Day weekend, and the Fourth of July occurred despite the fact that the United States had become the world epicenter of COVID-19. The people who did not comply with small sacrifices and did not hold the safety of others as a priority are responsible for the United States' exceedingly high number of cases. If these people had done something as simple as staying six feet apart and wearing a piece of cloth over their face, hundreds of thousands of lives would have been saved. The same thing can be said of simple, important actions that can reduce gun violence or climate change.

Injustice and Action

The pandemic highlighted many other deep-rooted issues. A disproportionate number of people of color (POC) were killed by the coronavirus. In POC communities, the resources necessary to fight COVID-19 were not distributed. One of my very good friends is an undocumented immigrant, and I could not believe my ears when he told me about the lack of testing facilities or easily accessible treatment in his neighborhood. Three of his aunts had COVID-19, and when they needed medical help, it was nowhere to be found. Our communities were hit hard by the pandemic in other ways. Small businesses, workers, and many others were drastically impacted by slow business and unemployment.

Unity is powerful, and we must work together to "build back better." Traumatic events bring about large changes in the attitudes of people. The collective suffering we have faced as individuals and as a species will be forever etched in our minds, but it has also spurred a movement for change in the ways that we live.

The pandemic has forced our government to reach across the aisle to create solutions to help those suffering from one of our country's worst economic crises. This cooperation allowed for stimulus checks to be doled out to millions of Americans in an effort to lessen the blows of the economic crisis. It is unfortunate that it took a catastrophic pandemic for government officials to work together. Yet we see that government cooperation is possible and necessary to defeat COVID-19. A new sense of American unity has been adopted by many, but not all, and the pandemic has shown the cracks in our foundation. We cannot let our country's polarization destroy us. Our country's delayed response to the pandemic helped cause the entirely preventable deaths of hundreds of thousands of Americans.

Climate Change

Climate change is like a pandemic; it's a less tangible crisis that requires large-scale cooperation. We cannot treat the climate crisis the same way that we treated the COVID-19 pandemic. We must take action.

2020 was also the fiftieth anniversary of Earth Day. In 2019, youth organizers started to plan massive strikes all over the country that would push government officials to take proper action against the climate crisis. Shortly after world-wide COVID-19 quarantines began, air quality improved in several cities and carbon emissions drastically reduced. We could no longer deny our climate impact on the planet.

Youth organizers had hoped the COVID-19 quarantine would end before April 22nd. A month before Earth Day 2020, we had to accept that what we had planned was not going to happen. I remember my heart dropping when I read the email from the coalition of US-based youth climate organizations announcing the cancellation. All plans to help revive a dying movement filled with conflict and infighting were scrapped, but we refused to let the pandemic stop us.

The youth climate movement is in every country and every city around the world and, as with COVID-19, technology connected us. We used our most powerful tool to our advantage. After taking a deep breath, we threw ourselves into the great undertaking of creating a massive, seventy-two-hour live stream Earth Day Live event. In under a month, our vision came to life. Solidarity and success came from all over the world as millions of viewers watched the live stream.

The fiftieth anniversary of Earth Day was not what we had pictured it to be, and yet, even though we did not have as much traction as we had hoped, we still accomplished an incredible feat.

Where We Stand Today

COVID-19 threw the world into chaos, displaying corruption, selfishness, and a lack of unity, but we overcame what was thrown at us. We had to adapt and become more creative, and we did. We saw the many faults of our systems, and we can no longer be silent about our country's inequality or wait to fight the climate crisis. We cannot let the divisions created by a cowardly fearmonger lead to our downfall. Through the trials and the tribulations, and despite a few outliers, we managed to help each other and come together. If we can do that with COVID-19, we must do it with climate change. We are capable of change and a better world.

The 2020 elections are paramount. We need the young and the old, the strong and the weak, anyone with the right to vote to go to the polls. Now is not the time to be apathetic. The future of our planet is at risk, and we will have to deal with the repercussions of the pandemic.

I cannot vote because I am only sixteen, but I am praying that on November 5th a new leader will be announced. I still hold hope that the United States will one day be the land of the free and the home of the brave. You must vote to protect the native inhabitants and caretakers of this stolen land. You must vote to protect the ideal of the American Dream that many immigrants risk their lives to achieve. You must vote to destroy the systems of oppression in our country. You must vote to protect our home. You must vote because our futures depend on it. So, I beg of you, cast your ballots so that my children and their children's children can have a future.

CHANGE

B-15Y Iceberg Antarctica No. 2 **by Zaria Forman.**
Soft pastel on paper, 60 x 90 inches

7

In Search of a Politics of Resilience
On the Field and in the Mind

Michael Pollan

Adapted from an interview conducted on July 23, 2020

THE FIRST TEACHABLE MOMENT of the pandemic, for me, had to do with what I saw happening in the supply chain. Early on, we had shortages developing in the supermarkets. The shortages did not just involve food; other everyday things were also in short supply. The first example people noticed was toilet paper. The big toilet paper shortage kind of mystified people, and the initial thinking was that people must be hoarding. Indeed, there was a certain amount of hoarding, but the shortage far exceeded that.

Eventually it came out that we have two distinct supply chains in America for toilet paper—as we do for food. One of them brings small packages of high-quality toilet paper to your local supermarket, while the other brings huge single-ply rolls to institutions, such as to the offices where we work, schools, prisons, and other commercial customers. When everyone began staying home because of the pandemic, they were using more toilet paper from the first supply chain and none from the second.

Over the years, we had developed a highly specialized but also very brittle supply chain system for toilet paper. A certain set of companies, dominated by a few giants, were manufacturing single-ply toilet paper in big rolls, which they sold and sent out to institutions. A different set of companies, making a slightly different product, were supplying the supermarkets. As one of these supply chains collapsed, the other came under incredible pressure. It's an illuminating example of how specialized the everyday products we use have become, and how dominated their markets are by a small number of companies. You might think you could have a company just say, "I'm going to move all this toilet paper over from these institutions to the supermarkets." But it was a different product, being made and sold in different sizes, and the companies just couldn't cope.

Brittleness, Efficiency, and Resilience

A system that is hyperspecialized is more vulnerable. Efficiency is a wonderful thing. It can result in lower prices, a better use of resources, and many other things. But there's always a tradeoff with efficiency, and that, of course, is the loss of resilience.

In times of crisis, resilience counts for more than efficiency. On television we've seen images of farmers spilling out tens of thousands of gallons of milk, or burying giant crops of onions or carrots. Those tended to be the farmers, ranchers, and dairy producers who were supplying the industrial, institutional food chain. Like our toilet paper supply, the milk supply chain that closed down was the one going to institutions—the giant boxes of milk that were being trucked to schools for those dispensing machines you see in cafeterias. It was sold very cheaply because of its large scale and the efficiency of the system. Suddenly, we needed a more agile supply chain that could take all that milk, put it into smaller containers, and reroute it to consumers in supermarkets, and we couldn't do it. Whatever gains in efficiency we'd created had very negative consequences in terms of the system's resilience. When COVID-19 came, we discovered that the system was fragile, rigid, and vulnerable.

A really important lesson to bring out of this crisis is *diversify to whatever extent you can*, even though it will cost you in some incremental way in efficiency. This lesson applies both to what you grow and who you sell to; it applies to the entire food chain. For a farmer it's more work to have ten crops instead of one. You may need a different kind of machinery to pull your carrots than you do to combine your wheat. There is a cost to diversity, but it gives you a cushion that's really, really important.

The principle that diversity is closely linked to resilience applies way beyond the pandemic. We have seen that more diverse farms do better in the aftermath of destructive climate events such as storms or droughts.

There have been interesting studies done in Central America comparing how different kinds of farms respond to hurricanes, showing which ones can withstand the storms and recover faster. Monocultures are vulnerable, but polycultures are resilient. This goes for both large and small farms.

I have seen this in my own community as well. There are two kinds of farms: the large mono-crop farms, which are highly efficient and often supply the institutional food system, and the diversified farms. Whether they are organic or industrial, these tend to be smaller and grow many different crops. The diversified farms have their own dependencies; for example, under normal circumstances, they tend to be dependent on restaurants. Restaurants have driven the farm-to-table movement and have done wonders for farmers trying to grow a diversity of high-quality crops. Suddenly, after COVID-19, the restaurants were closed, but these farmers were able to pivot fairly quickly. The reason, I think, was that they hadn't lost touch with eaters and what eaters desired. Many of them were able to move to a community-supported agriculture (CSA) model, selling produce in boxes where they could decide the composition of the boxes each week. As CSA customers know, these boxes have a great diversity of things in them and vary over time as the seasons change.

These farmers are doing quite well right now. Even though there are inefficiencies in terms of their size and the number of different crops they grow, their lower level of mechanization has allowed them to move from filling crates for restaurants to filling boxes with a diversity of produce directly for consumers. This has been true even for milk. Farmers with small pasteurization operations have pivoted to selling milk as a finished product to individual people who come to the farm to pick it up.

The Consumer versus the Citizen

Business doesn't care about long-term vulnerability. It cares about short-term profits. Companies have been able to show that consumers benefit from economies of scale—and in some ways, that is true. Meat prices, for instance, relative to other goods in your shopping cart, have gone down in the last fifty years. Meatpacking companies have such scale that they can force down the prices they pay to their suppliers. As a result, meat is very cheap in America, and to many people this is seen as a great blessing. But we're now paying the cost of that worship of efficiency.

What's good for a consumer's buying power and pocketbook is not necessarily good for people as moral and biological beings. Being a consumer is only one of the many identities we each have. But in a capitalist culture, the consumer identity—where someone looks for good value and low prices above all else—gets the most attention. A culture that celebrates our identity as consumers above all else is responsible for our highly vulnerable supply chains today; it's also the main cause for the collapse of antitrust enforcement in our country.

Antitrust law was initiated in the first decades of the twentieth century, during the first Roosevelt administration, to protect our society against overconcentration of corporate power. At that time there was a general understanding that these "trusts," as they were called, were a peril to society, specifically to the polity. It was recognized as undesirable to have accumulations of power so great that they could push the government around, so we went about reducing the size of those companies by passing the Sherman Antitrust Act and a couple of other laws. We broke up the meat industry, for instance, because we determined it was a bad idea to have five companies slaughtering eighty percent of the beef.

Something important happened to antitrust enforcement during the Reagan administration. Antitrust law was redefined by a policy memo in the Justice Department that was heavily influenced by the theories of Robert Bork, the jurist who famously lost his bid to be a Supreme Court justice. In 1978, Bork published a book, *The Antitrust Paradox*, arguing that we should only worry about the size of corporations when mergers or combinations hurt the consumer.[1] That is, if we could not show damage to the consumer—specifically, escalating prices—then there was no reason to stand in the way of mergers. That was a curious standard because it had nothing to do with the foundation or purpose of the antitrust laws. The Sherman Antitrust Act is about concentrations of corporate power and how they can hurt producers. Its concern is with smaller producers in a supply chain having no negotiating power against a handful of enormous companies, such as the ranchers confronting the meat packing companies. The Act makes no mention of the consumer. But in the neoliberal political climate of the 1980s, this new consumer-centric idea seemed sensible to a lot of people in both parties. The Reagan-era advisory memo has become antitrust policy in America, ushering in a wave of consolidation and mergers that continues to this day. As a result, we have a situation that's worse than when the trustbusters began their work: now there are only *four* meat packers slaughtering eighty percent of the beef. Incredibly, all it would take is a new memo to change things, but that hasn't happened.

1. Bork, *Antitrust Paradox*.

Habits of Thought

In this pandemic moment, some habits are being broken and others are being reinforced or built for the first time. As someone who travels extensively to speak and to do reporting, it's been a really radical shift in my life not to be going anywhere. I've had moments of feeling really restless, and the fact that every day is much like the one before sometimes takes an emotional toll. It's made me think a lot about how important it is to change our minds, rather than our locations.

One of the things that's changed personally for me is that I've pursued my meditation practice with a lot more commitment—and a lot more success—because that interior mental space is all I have for travel. My meditation teacher has been teaching with a quote from (Shakespeare's) Hamlet, who also suffered from a kind of psychological claustrophobia: "I could be bounded in a nutshell and count myself the king of infinite space." So, this is the challenge for us: here we find ourselves in our nutshells, in a contracted world. How can we turn ourselves into kings of infinite space?

So much of American history has been about physical movement: finding the next place where you can escape history (and other people who've oppressed or annoyed you) and start out again. But there's this sense now that we really are on a spaceship, and in enclosed spaces you have to be careful. The virus is fostering a commonality that we weren't aware of, a biological commonality, where someone else's health and behavior can really affect our own—whether they have health insurance, whether they are wearing their mask. It's a rehearsal for climate change in that we recognize that there is no escaping our predicament. There's no other place to go. We're coming to see that we're really all in the same boat.

This is a painful awakening in some ways. I think we'd rather be able to go to that next frontier (and that fantasy is still alive—Elon Musk and Jeff Bezos are still feeding frontier fantasies about interplanetary travel). But for most of us it's a bit like that Apollo 8 moment, when we saw those first photographs of the Earth from space. Because it created the sense of a shared habitat, that moment changed everything. Like those images, the pandemic has made the world seem smaller.

Much like our food systems, we've organized our mental life around the ideal of efficiency. The goal is to get things done. We depend upon things like travel and new information to change our habits and the contents of our minds, but we tend to overlook the fact that it is within the powers of consciousness to effect those changes. To connect to what I wrote about in my most recent book,[2] one of the things you learn when you start exploring psychedelics is that there are many types of consciousness. There is ordinary consciousness, consciousness linked to meditation, consciousness linked to psychedelics (and even there, different psychedelics can produce different types of consciousness). This awareness of consciousness as plural makes me think we should avoid monocultures of consciousness, just as we should avoid monocultures in the field and in our supply chains. There's a resilience in mental diversity, too. These other states of consciousness can give us the perspective that things don't have to be the way they are, that they could be different. We need to cultivate this kind of resilience in place of what we have been cultivating, which is optimization.

We tend to get good at what we can measure, and this is a problem. Resilience cannot be easily measured. Part of the attraction of efficiency is that you can put numbers on it, but resilience calls for another kind of economics. That's probably why we haven't yet gotten very good at resilience, and why we haven't appreciated its importance.

A New Politics

COVID-19 has exposed many things about the way we live. Even though it's only been a few months, I'm optimistic we can build a new politics based upon what we have learned—a politics that really connects the dots. Already we know what needs to change to increase security, to make the systems we depend on more resilient and more just.

One habit of thought that the pandemic has helped us break out of is the idea that we are separate atoms and that we can protect ourselves with money and geography. When New York was having its COVID-19 crisis, people elsewhere in the country gloated that such things only happened in blue states, that red states were somehow protected. There was this habitual desire to draw lines of defense. Yet every one of those lines has broken down.

Another habit of thought has been broken by a revolutionary new idea: the "essential worker." This concept includes healthcare workers, childcare workers, schoolteachers, food workers, people in meat plants, people picking our food, people serving our food. We have tended to pay people who do very important work very little. Yet suddenly, during the pandemic, they have become recognized as "frontline workers" who are essential to our continued existence—a massive departure from how most of us thought about these people just months ago. While those of us who made really good incomes could stay home to work and remain safe from the virus, these

2. Pollan, *How to Change Your Mind.*

other people on whom society absolutely depended had to return to their workplaces. It turned out we were very reliant on these people in ways we hadn't appreciated. That moment was driven home when President Trump forced the meat plant workers to go back to work, despite the fact that doing so was enormously dangerous to the workers. Could there be a stronger argument here for paying people who do work we depend on a living wage? The politics here are shifting. We don't want to miss this opportunity, in the same way that we don't want to let pass the strong case for universal health insurance and for recognizing the absurdity of tying your health insurance to your employment. The quality of your healthcare influences my health, and vice versa. This has never been more clear.

A third habit of thought we need to break out of is our tendency to lose track of complex societal connections. One of the really striking things I've learned in years of covering the food system is that there are farmers and companies who lose track of who's eating their food because the food supply chain is so long and opaque. A famous example is Monsanto: they ran into so much trouble marketing genetically modified crops because they lost sight of the fact that there were humans—eaters!—at the far end of the system they were selling seeds into. They thought that they were just talking to farmers about efficiency and yields and the convenience of having a crop that was resistant to herbicide. But they completely lost track of the fact that people were going to have to be willing to eat the stuff, and these people might have some reservations about eating a genetically modified crop. I have seen this phenomenon over and over again, with ranchers, with people operating feedlots. They're all working for the next person up the chain and have lost track of the chain in its entirety. This creates a certain sloppiness, which leads to systems that tolerate high food safety risks, among other problems. It's a casualty of a complex industrial economy. You're much more scrupulous if you remember you're actually feeding somebody at the end of the day. Smaller farms have more contact with the people they feed and so tend not to fall into that trap.

The same abstraction can occur in other areas of life as well. Most Americans have treated politics as a spectator sport—to the extent that they pay attention to it at all. The idea that decisions made in the White House would directly affect our lives had been invisible to many people. Not any more: because of the mismanagement of the pandemic by the Trump administration, our lives have been upended. The last time I can remember feeling such a direct line between the President's actions and my own life was during the Vietnam War, when we had the draft. Since then we have been cushioned from such obvious cases of political cause-and-effect. Most of the time, you have to go through a complicated and abstract process to figure out how, say, a Fed decision affects your ability to buy a house. The fact that politics matter deeply and affect all of us has been driven home by the pandemic, and this recognition will shape this election in the fall.

The pandemic has taught us the importance of paying attention. Now we know that we need to pay attention to how things work, how our food system is organized, who's in the White House, how we pass our time, how we spend our mental energy. I feel like we've been sleepwalking. Suddenly the present moment demands our full attention. This is the first step toward driving change: paying attention and realizing that change is necessary. The Black Lives Matter movement is a sign of that awakening. We've been sleepwalking through the tragedy of racism—or at least white people have been—and suddenly we can't do that anymore.

We're often told that crises are opportunities. The first step in making good use of this crisis is waking up and paying attention. Everything after that gets really complicated, because now we enter the messy realm of politics. But I think that there are opportunities for change, that we can built a new politics on what the wreckage of the pandemic has forced us to see. I can't tell you what shape this new politics will take, but I'm encouraged that we are at least waking up, and that's the first necessary step.

References

Bork, Robert H. *The Antitrust Paradox: A Policy at War with Itself.* New York: Maxwell Macmillan International, 1993.

Pollan, Michael. *How to Change Your Mind: What the New Science of Psychedelics Teaches Us About Consciousness, Dying, Addiction, Depression, and Transcendence.* New York: Penguin, 2018.

8

Caring for People and Nature First
Four Cornerstones for a Successful
Progressive Agenda

RIANE EISLER

LIKE FAMILIES, POLITICS, EDUCATION, and religion, economies don't spring up out of thin air. Economies are human creations, governed by the values and worldview of the larger society of which they are a part. Einstein famously noted that we cannot solve problems with the same thinking that created them. In the following few pages, I want to examine societies and economies that go beyond the familiar categories: capitalist/socialist, right/left, religious/secular, Eastern/Western, Northern/Southern. In their place, I would like to explore the new categories of *the partnership system* and *the domination system*. As we begin to use these new categories, we learn to identify what is needed for a successful progressive agenda and to recognize constructive actions that each of us can take.

Old Thinking

If we employ the partnership-domination social scale, we see that Adam Smith and Karl Marx constructed their capitalist and socialist economic theories as responses to an age that still stood much closer to the domination side. Both capitalism and socialism were attempts to move away from the domination economics that pervaded most of recorded history. Examples included the top-down economics of tribal chiefs, Chinese emperors, Middle Eastern sheiks, and European kings and feudal lords. Adam Smith was interested in challenging the system of *mercantilism*: the control of economics from the top by kings and court officials. Marx challenged capitalism and the exploitation of workers and peasants by so-called nobles and the growing bourgeoisie.

While many today focus on the differences between Smith and Marx, both men actually saw nature as something to be dominated and exploited. They didn't consider the damage to our natural life-support systems or the need to care for them. Smith advocated unlimited economic growth guided by pure self-interest and by the "invisible hand" of the market, and Marx envisioned unlimited industrial expansion controlled by the needs of the proletariat. Both assumed the domination of nature in their systems.

As for the life-sustaining work in households—caring for children, protecting people's health, maintaining a clean and healthy home environment—these were simply not considerations within their economic equations. For both Smith and Marx, these vital activities were just "reproductive" rather than "productive" work. Neither recognized the economic value of the work of caring for people and establishing an environment within the home that fosters health and safety. To them this "women's work" was invisible, something to be performed for free in male-controlled households. As late as the mid-nineteenth century when Marx was writing, women's work both in homes and in the market was legally the property of their fathers or husbands. If a woman was negligently injured, she could not sue the wrongdoer for her injuries. Only her husband could—as compensation for his loss of her services.

Today, at least in most regions of the world, women are no longer the property of their fathers or husbands. But the devaluation of care work is still the economic norm. This devaluation is part of the gendered system of values that underlies both capitalism and socialism. According to the US Department of Labor, the pay for caregivers in the market is woefully low; childcare workers in the United States earn less than dog walkers. In most regions of the world, women are still supposed to perform this work for free in homes.

COVID-19 has laid bare the structural faults in our economy and society. We see the systemic failure, inequity, and unsustainability of the world around us. "Essential workers" have been charged with the duties of caring for people, providing them with healthcare, food, a clean environment, and other necessities. Yet this work is paid little in the market and is not generally supported by government policies.

How is it that such injustices remain? Can we build a different world with a more just economy and society? What will it take to reach this goal?

New Thinking

If we are going to build an economy to meet the challenges we currently face, we will need a new economic map. At present, the field of economics as it is generally taught in universities and textbooks only includes the market, government, and illegal economic sectors. The standard models often ignore the economic contributions of three life-sustaining sectors: the natural economy, the unpaid community economy, and the household economy.

It is essential for a new economic map to include these vital sectors of the economy. I could argue for this conclusion for human or environmental reasons, but it is also *economically* advantageous. In our knowledge-service age, economists keep telling us, the most important resource is "high quality human capital." We know from both psychology and neuroscience that high quality capital is largely dependent on the quality of the care and education that children receive early on in their development. The formation of resilient, flexible people who learn to work with others and to adapt quickly to new circumstances (including new technologies) hinges on unaccounted-for labor. In other words, it hinges on the so-called "women's work" that is not valued in either capitalist or socialist theories as productive work.

My book, *The Real Wealth of Nations: Creating a Caring Economics*, calls for a caring economy of *partnerism*.[1] This kind of economy adequately rewards caring for people through economic rules, practices, and policies. Work that starts in early childhood is often invisible to our current economic valuations. My most recent book, *Nurturing Our Humanity: How Domination and Partnership Shape Our Brains, Lives, and Future*, describes four cornerstones that are necessary in order to shift from a system based on domination to one of partnership.[2]

The good news is that we do not start from square one. If we look at modern history through the lens of the domination-partnership scale, we can trace how one progressive social movement after another has challenged the same thing: a tradition of domination.

The eighteenth-century Enlightenment "rights of man" movement challenged the *divinely ordained right of kings* to rule their subjects. The

1. Eisler, *Real Wealth of Nations*.
2. Eisler and Fry, *Nurturing Our Humanity*.

nineteenth and twentieth-century feminist movements challenged the *divinely ordained right of men* to rule women and children in the castles of their homes. Abolitionist, civil rights, anti-colonial, and Black Lives Matter movements challenged the *divinely ordained right of a superior race* to rule inferior ones. Economic and social justice movements challenge top-down economic and social control. Pacifist and peace movements—and more recently the movement to end the pandemic of violence against women and children—challenge the use of force to impose and maintain forms of domination. The environmental movement challenges yet another tradition of domination that barrels toward an evolutionary dead end: the conquest of Nature.

As a result of these partnership movements, we see the shift from monarchies to republics, the outlawing of torture, and the birth of human rights. Colonialism dwindled, racism and anti-Semitism are increasingly condemned, and nonviolent regime changes now outnumber violent revolutions by a ratio of two-to-one.

However—and I want to emphasize this point—challenges to top-down control have focused primarily on domination in politics and economics. If domination is thought of as a pyramid, this is the top. The majority of progressive social movements pay scant attention to the relations that most profoundly affect how our brains develop. I am talking about parent–child and gender relations. Consequently, the domination pyramid continued to rebuild on the same foundations in regression after regression. This happened through totalitarianism, religious fundamentalism, and other kinds of domination-isms.

Four Cornerstones for Fundamental Social and Economic Change

If we wanted to build a house, we would start with some kind of plan for constructing its foundations. Likewise, successfully building a more equitable, caring, and sustainable economy and society requires changing four interconnected cornerstones. These changes address the foundations of the domination pyramid in our efforts to support partnership.

The First Cornerstone: Childhood

Neuroscience shows that the neural pathways of our brains are not permanently set at birth. As I showed in *Nurturing Our Humanity*, they are largely formed through interactions in a child's early experiences and observations. Therefore, partnership-oriented families that model mutually respectful, empathic, and egalitarian relations are an essential foundation of partnership societies that are oriented to equity, sustainability, and justice.

I am not equating parenting with a laissez faire approach; parenting should be *authoritative* but not *authoritarian*. This means that, instead of hierarchies of domination girded by fear and force, there are hierarchies of actualization. Here respect and accountability flow not just from the bottom up but are rooted in mutuality. Power is *power with* and *power to*, rather than *power over*.

When children see and experience rankings of domination grounded in fear and force, it animates their mental model for all relations. This is why regressive political agendas put so much emphasis on teaching children—before their critical faculties are formed—that an authoritarian, male-headed, punitive family is either divinely or biologically ordained.

Some nations have shown that it's possible to change habits of domination parenting through education, laws, and media. Starting with parenting, and from preschool to graduate school, we need to embrace an education of caring relations. An easy to use resource is the "Caring and Connected Parenting Guide."[3] We must encourage leaders (including spiritual and religious leaders) to take a strong stance against intimate violence. Every year these acts of violence take the lives of millions of women and children worldwide—they provide models of violence in all relations.

The Second Cornerstone: Gender

We typically don't think of gender or childhood as essential to a successful progressive agenda, but this must change. How a society constructs the roles and relations of these two human forms—male and female—not only affects each individual's life options; it also affects families, education, religion, politics, and economics.

It's not coincidental that regressive regimes impose or maintain an authoritarian, male-dominated, highly punitive family structure. We see this across many regimes throughout history: rightist, secular, and Western, as in Hitler's Germany; leftist, secular, and Western, as in Stalin's USSR; religious and Eastern, as in ISIS, Iran, and the Taliban; or religious and Western, as in the rightist-fundamentalist Christian alliance in the US. Studies show that highly prejudiced people who respond to scapegoating and misogyny, and

---

3. Center for Partnership Studies, "Homepage."

who vote for strongman leaders, typically grew up in authoritarian, rigidly male-dominated, and highly punitive families.

Domination systems are based on in-group/out-group thinking. We see this in the in-group of "mankind" and the out-group female "other." Not only does it create an "other," this male/masculine and female/feminine paradigm also prepares people to equate difference with superiority or inferiority. Using this template, you locate your identity at one pole or the other: domination or being dominated, serving or being served, for example. This same dichotomous thinking expands to racism, religious differences, and ethnic differences. President Trump clearly articulated this worldview when he berated governors and mayors who did not use force against protestors after George Floyd's killing. He called them weak and declared that it's all about domination. People who acquire this mindset see only two possibilities: you either dominate, or you are dominated.

The domination system's subordination of women and "the feminine" directly impacts society's guiding values. This of course includes what we value in economic terms. This dynamic would seem obvious if we weren't so used to marginalizing anything we've been taught to think of as a "women's issue." Along with the subordination of women in domination systems comes the subordination of traits and activities that are stereotypically associated with femininity. In systems of domination these traits are deemed unfit for "real men."

This gendered system subordinates the values of caring, caregiving, and nonviolence. Unsurprisingly, this system of subordination translates directly into our social and fiscal priorities. In places where women are devalued, one finds higher investments in "hard" or "masculine" priorities like prisons, weapons, and war, where domination is imposed and maintained. In countries where women are more highly valued—as I will illustrate below—there is higher investment in caring for humans and nature. In these contexts, caring is not devalued or seen as inferior.

The Third Cornerstone: Economics

Once we look at economics from a gender-holistic perspective, we see that the gendered system of values adversely affects a society's general quality of life. Based on statistical data from eighty-nine nations, the 1995 Center for Partnership Studies' report *Women, Men, and the Global Quality of Life* showed how the status of women is actually a powerful predictor for our general quality of life. Since this study, others like the World Values Survey and the World Economic Forum's Gender Gap Reports have confirmed this connection as well. We cannot ignore the relationship between the status of women and a nation's economic success and quality of life.

Yet these findings are frequently dismissed—not only by economics departments and texts, but by the media. The dismissal continues despite the highly visible relationship between gender equity and value systems in nations like Sweden, Finland, and Norway. These countries, which used to suffer famine and poverty, regularly score high in the World Economic Forum's Global Competitiveness reports. They have the lowest gender gaps (forty to fifty percent of national legislators are female), low crime rates, high longevity scores, and sit at the top of international happiness reports.

Of course, these societies are not perfect; still, they don't have huge gaps between the haves and the have-nots. These countries are also not socialist societies in the sense that they operate very healthy market economies. Their more equitable distribution of resources is not because they are relatively small and homogeneous—there are lots of relatively small and homogenous nations that are very domination-oriented. Instead, these countries are proudly shaped by the values of caring societies. Women's rise in status was accompanied by an increase in caring policies. As this shift occurred, these societies moved toward the partnership side of the partnership-domination scale.

A major reason for these nations' ascent from dire poverty to prosperity is that they pioneered caring policies like universal healthcare, high quality early childhood education, generous paid parental leave, and elder care with dignity. They also worked to abandon traditions of violence, for example by pioneering the first Peace Studies programs and the first laws against physical discipline in families. And, as you might expect, they are rapidly shifting to solar and other renewable energy sources to extend these values of care toward Nature as well.

In short, they are really helpful examples of what it looks like for us to invest heavily in our human and our natural infrastructure. These changes are possible. We can move away from systems of domination and toward a caring economics of partnerism.

The Fourth Cornerstone: Narratives and Language

Social psychologists have long told us that the categories provided by a society's language channel people's thinking. This makes it very difficult for them to see any alternatives. Our new language of the *partnership system* and the *domination system* addresses this need. But we also need new stories. We need more accurate stories about our human possibilities that are based on evidence from the social and biological sciences. In *Nurturing Our Humanity*, I describe studies that show how the pleasure centers in our brains light up far more when we share and care than when we win.

There is also mounting archeological and ethnographic evidence that a durable legacy of cultural partnership lasted for tens of thousands of years. Partnership first appeared when humans lived as foragers, and then again in early farming settlements such as Catal Huyuk and Bronze Age civilizations such as Minoan Crete. But during a time of great disequilibrium between five and ten thousand years ago, most world regions shifted from partnership to domination.[4]

Nevertheless, even after this shift there were intermittent modes of partnership that resurged. For example, the teachings of Jesus and early Christian communities (or, more accurately, Jewish communities where Jesus preached) were more oriented toward the partnership side of the scale. We read in the official Christian scriptures that women played important leadership roles in the early church, and Mary Magdalene—dismissed in these writings as a whore—was first to see the risen Christ. Moreover, in the Gnostic gospels, early non-canonical sources on the life and teachings of Jesus, Mary Magdalene is described as Jesus's partner—not a whore.

In more recent centuries, during the transition from the agrarian to the industrial age, we begin to see strong movement toward a second major shift, this time from systems of domination to the partnership side of the social scale. Unfortunately, as I mentioned above, the movement to leave behind traditions of domination marginalized relations for the majority of humanity: women and children.

For many of us who consider ourselves progressive, concerns for women and children have been secondary. Our "education" has continued to marginalize women and children in both our conscious and unconscious minds. Only as we become more aware of the vastness of this exclusion can we formulate and carry out an effective and progressive agenda that will change these narratives.

In the present moment, we need new economic narratives that recognize the value of care work. We need to take steps toward partnerism that value caring for others, social justice work, and caring for the environment. These values require new metrics that—unlike GDP and most current proposals for GDP alternatives—reflect the economic value of caring for people and nature.

In 2014, the Center for Partnership Studies developed a prototype: twenty-four Social Wealth Economic Indicators that take into account the

---

4. See Eisler, *Chalice and the Blade*; and Eisler, *Sacred Pleasure*.

impact of childhood care and education on human capacity development. They pay attention to women and children and recognize their crucial role in reducing poverty. They also show how other "out-groups"—different races and ethnicities who provide care work—are paid poverty wages. Many of the women who do this work in the US, mostly Latinas and African Americans, have to rely on welfare so that they and their families can survive. Currently, these metrics are being updated and condensed into an easily accessible Social Wealth Index (SWI).[5]

Conclusion

The movement toward a partnership-oriented world has not failed, but it is incomplete. We all have the opportunity to contribute to this movement by expanding the progressive agenda. We can all focus on the four cornerstones—childhood, gender, new economics of partnerism, and new language and stories—to more accurately reflect our past, present, and the possibilities for our future together.

Human societies and economies are human creations. Together we can lay the missing foundations for a more sustainable and equitable world. Every one of us can play a part in accelerating the shift from domination to partnership even as we navigate the challenges of COVID-19.

References

Center for Partnership Studies. "Homepage." Accessed August 8, 2020. https://centerforpartnership.org.

Center for Partnership Studies. "The Social Weath Index." Accessed August 8, 2020. https://centerforpartnership.org/programs/caring-economy/social-wealth-index.

Eisler, Riane, and Douglas P. Fry. *Nurturing Our Humanity: How Domination and Partnership Shape Our Brains, Lives, and Future.* New York: Oxford University Press, 2019.

Eisler, Riane Tennenhaus. *The Chalice and the Blade: Our History, Our Future.* Cambridge: Harper & Row, 1987.

———. *The Real Wealth of Nations: Creating a Caring Economics.* San Francisco: Berrett-Koehler, 2007.

———. *Sacred Pleasure: Sex, Myth, and the Politics of the Body.* San Francisco: Harper, 1995.

5. Center for Partnership Studies, "The Social Weath Index."

9

Re-aligning Technology with Humanity

Tristan Harris

Adapted from an interview conducted in July 2020

MODERN TECHNOLOGY HAS CREATED an attention economy by hacking our instincts to get us hooked to our screens. From Facebook to Instagram, TikTok to Tinder, the digital world is where we go to find everything—from people to date, to products to buy, to magazines, books, or other popular culture that help us understand our world. We live in a digital ecology where our attention—our consciousness—has become a valuable but scarce commodity. When powerful tools hijack human attention, myriad social problems result, such as political polarization and technology addiction. While our global challenges are increasing exponentially, our ability to agree with each other, engage in critical discourse, and make good decisions is going down. It is possible to create a digital ecosystem that is in alignment with humanity and our values. But it will take some big shifts to get us there, and time is running short.

The Attention Economy Robs Us of Choice

Seeing reality clearly and truthfully is fundamental to our capacity to do anything. Facebook manages the attention flows and thoughts of more than three billion people—more people than the psychological footprint of Christianity. The digital ecology it dominates, enabled by overwhelmingly powerful artificial intelligence, is integrated into an economic system that places profit above all else. That profit is generated by advertising, which must be noticed in order to be successful. But there are only so many hours in the day that one can stare at a screen. And so a relentless competition for our attention has emerged that drives what I call "the race to the bottom of the brainstem": the process of hacking more and deeper dimensions of our psychological limits, biases, and emotions in order to keep us hooked to enticing online platforms.

Choice springs from the minds and nervous systems of human beings. By monetizing attention, we've sold away our ability to see problems and our ability to collectively choose to change the logic of our economic system. People working to create a world that's prosperous for all of us are seeking the same thing—to not have our economies and the logic that drives them be built on separation, extraction, and commodification. When you maximize separation, you remove understanding and respect, the interconnections and relationships, that are the basis of our society. In food, you lose touch with the life cycle that makes agriculture sustainable. In education, you lose the interrelatedness of human development, trust, care, and teacherly authority. In love, you sever the complex terrain of human relationship when you turn people into playing cards on Tinder. And when you commodify communication as blocks of comment threads on Facebook, you remove trust, nuance, and respect. In all these cases, extractive systems slowly erode the foundations for a healthy society.

If you point an economic system based on separation and extraction at the Earth's resources, you have a little bit of time before you destroy our world. But when you point that same economic system at our attention, it quickly destroys our ability to make choices. Choices are what allow us to do something different than what is hurting us. But choice depends on discovering that there's an alternative to what seems to be inevitable. News feeds on Facebook or Twitter operate on a business model of commodifying the attention of billions of people per day, sorting for what tweets, posts, or groups get the most engagement (clicks, views, and shares)—what gets the strongest emotional reactions. Because of this, the attention economy generally trades on the negative—on salacious stories and stories of conflict—and not on the positive. When a positive development happens but doesn't evoke strong powerful emotions, we don't see or hear of it and therefore don't pay attention to it. This negative lens of reality has operated the collective beliefs, biases, and behavior of three billion people over the last decade.

These commodifying attention platforms have warped the collective psyche. They have led to crazier views of the world. YouTube's recommendation algorithms, which determine seventy percent of daily watch time for billions of people, "suggest" similar videos but actually drive viewers to more extreme, more negative, or more conspiratorial content because that's what keeps viewers on their screens longer. For years, YouTube recommended "thinspiration," anorexia-promoting videos to teen girls who watched videos for "dieting." And when a person watched science videos of NASA's moon landing, YouTube recommended videos about the Flat Earth conspiracy. They did this hundreds of millions of times. News feeds and recommendation systems like this have created a downward spiral of negativity and paranoia, slowly decoupling billions of people's perception of reality from reality itself.

Shifting Systems to Protect Attention

E. O. Wilson, the famed biologist, proposed that humans should run only half the Earth, and that the rest of our world should be left alone. Imagine something similar for the attention economy. We can and should say that we want to protect human attention, even if it sacrifices a portion of the profits of Apple, Google, Facebook, and other large technology corporations.

Ad blockers on digital devices are an interesting example of what could become a structural shift in the digital world. Are ad blockers a human right? If everybody could block ads on Facebook, Google, and websites, the internet would not be able to fund itself, and the advertising economy would lose massive amounts of revenue. Does that outcome negate the right? Is your attention a right? Do you own it? Should we put a price on it? Selling human organs or enslaved people can meet a demand and be profitable, but we say these items do not belong in the marketplace. Like human beings and their organs, should human attention be something money can't buy?

The COVID-19 pandemic, Black Lives Matter movement, climate change, and other ecological crises have made more and more people aware of how broken our economic and social systems are. But we are not getting to the roots of these interconnected crises. We're falling for interventions that feel like the right answer but instead are traps that surreptitiously maintain the status quo. Slightly better police practices and body cameras do not prevent police misconduct. Buying a Prius or Tesla isn't enough to really bring down levels of carbon in the atmosphere. Replacing plastic straws with biodegradable ones is not going to save the oceans. Instagram, with its constant social comparison and signaling, may be systemically hijacking the human drive for connection, but hiding the number of "Likes" on Instagram is not going to solve teenagers' mental health problems. We need much deeper systemic reform. We need to shift institutions to serve the public interest in ways that are commensurate with the nature and scale of the challenges we face.

Here at the Center for Humane Technology, one thing we did was convince Apple, Google, and Facebook to go against their economic interests and adopt—at least in part—the mission of Time Well Spent. This was a movement we launched through broad public media awareness campaigns and advocacy that gained credence with technology designers, concerned parents, and students. It called for changing the digital world's incentives from a race for "time spent" on screens and apps into a "race to the top" to help people spend time well. It has led to real change for billions of people. Apple, for example, introduced "Screen Time" features in May 2018 that now ship with all iPhones, iPads, and devices. Besides showing each person how much time they spend on their phone, Screen Time offers a dashboard of parental controls and app time-limits that show parents how much time their kids are spending online (and what they are doing). Google launched its similar Digital Wellbeing initiative around the same time. It includes further features we had suggested, such as making it easier to unplug before bed and limit notifications. Along the same lines, YouTube introduced "take a break" notifications to viewers.

These changes show that companies are willing to make small sacrifices, even in the realm of billions of dollars. Nonetheless, we have not yet changed the core extractive, separating, and commodifying logic of these corporations. Doing something against one's economic interest is one thing; doing something against the DNA of a company's purpose and goals is a different thing all together.

Working Towards Collective Action

We need deep, systemic reform that will shift technology corporations to serve the public interest first and foremost. We have to think bigger about how much systemic change might be possible, and how to harness the collective will of the people.

In the case of the tech industry, one head start is that we don't need to convince hundreds of countries or millions of people. Fewer than ten people run the twenty-first century's most powerful digital infrastructure—the FAANG stocks, representing the stocks of Facebook, Amazon, Apple, Netflix, and Alphabet (formerly Google). If those individuals got together and agreed that maximizing shareholder profit was no longer the common aim, the digital infrastructure could be different. Because of their asymmetric power, these individuals could make different choices and change their economic logic to work in humane and ethical ways.

Recently at the Center for Humane Technology, we interviewed Christiana Figueres, the Executive Secretary of the United Nations Convention on Climate Change (2010–2016), for our podcast. She was responsible for the "collaborative diplomacy" that led to the Paris Agreement, and we learned how she was able to do this—to get 195 different countries, against all odds, to make shared, good-faith goals towards addressing climate change. Christiana initially didn't believe it was possible to get that many countries to agree, but she realized that successfully hosting the Paris Convention meant that she herself would have to change. *She* had to genuinely believe it was possible to get the countries to commit to climate action. That was how she was able to then focus on getting the participating countries to believe in the possibility of addressing climate change as well. Where earlier international climate negotiations had failed, Christiana's efforts brought nations together to agree on financing, new technologies, and other tools to keep global temperature rise below 2 degrees Celsius and, even better, to limit the rise to 1.5 degrees Celsius. If Christiana could do this with 195 nations, we could consider the possibility of doing it with ten corporate heads.

Of course, leaving power concentrated in the hands of ten people is not healthy. It would not be a long-term solution. Like everyone else, tech leaders only have so much time and attention available to address a massively expanding set of issues. So a possible approach is to redistribute their power as a massively decentralized system, where every individual has concern for the whole. For instance, Facebook could hand over governance, design, and harm-reduction practices to citizen juries run by the people, using the procedures of what's come to be known as "liquid democracy." Is it feasible to distribute power in this way? Is it possible to have a decentralized group of people who want the right things, and in a way that doesn't commodify the basis of our society?

A few early answers may be found in the nascent realms of stakeholder-based ownership models, blockchain technology, new forms of democratic decision making, and distributed autonomous organizations (DAOs) that operate under a different economic logic. The tech industry could be held and managed more as a collective commons, putting decision making and power in the hands of the people who use the technology, and engaging the public in co-creating the directions developed. This is counter to the current oligarchic model, where the power is held in the hands of very few people, with little oversight, and poor alignment of incentives to serve the public interest.

A New Economics of Humane Technology

Several economic principles need to shift in order for technology to align with humanity and the planet. One of these is the growth paradigm. You simply can't run a logic of infinite growth on a finite substrate. The drive for infinite economic growth is leading to a planetary ecological crisis. For tech companies, the focus on infinite growth of human attention when only a finite amount of attention is possible leads to a similar crisis of global consciousness, wellbeing, and democracy. We need to shift to a post-growth attention economy that places mental health and wellbeing at the center of our desired outcomes. A small hint of this shift is taking place in countries such as New Zealand and Scotland, where organizations such as the Wellbeing Economy Alliance are working to shift from an economy that promotes the gross domestic product (GDP) to one that promotes wellbeing. Leaders are asking how wellbeing can inform public understanding of policies and political choices, guide decisions, and become a new foundation for economic thinking and practice.

Another shift toward a more humane technology requires having a broader array of stakeholders who can create accountability for the long-term social impact of our actions. Right now, it is possible for large technology companies to make money by selling thinner and thinner "fake" slices of attention—selling fake clicks from fake sources of fake news to fake advertisers. These companies make money even if what the link or article leads to is egregiously wrong and propagates misinformation. This opportunism debases the information ecology by destroying our capacity to have trusted sources of knowledge and shared beliefs in what is true, which in turn destroys our capacity for good decision making. The result is polarization, misinformation, and the breakdown of democratic citizenship. We need to create mechanisms that incentivize participants in the digital world to take into consideration longer time frames and the broader impact their actions are having on society.

Technology companies also need to shift away from the die-hard competition that undergirds profit maximization and the attention economy. Could there be a basis in the technology sector for working together towards shared common goals? Is hyper-competition necessary for the technology sector to operate and develop? How do you fundamentally change the basis of competition towards an orientation of mutuality and cooperation? It's difficult for any one actor to say, "We are going to have growth and profit simply inform rather than run our business," when other players are still playing within a landscape of competing for finite resources and power. Legislation and policy can certainly help, but these tools operate at a much slower pace than the rate that is needed to make a difference. The technology sector itself needs to come together and find ways to operate so that shared societal goals are placed above hyper-competition and profit maximization.

Human will plays an important role here. What if the leaders behind Apple's App Store revenue distribution model—which acts as the central bank or Federal Reserve of the attention economy—simply chose to distribute revenue to app makers based not on whose users bought the most virtual goods or spent the most time using the app, but on who among the app makers best cooperated with other apps on the phone to help each person in society live more by their values?

Ultimately it comes down to setting the right, fair rules of the game. It is difficult for any one actor to optimize for wellbeing and alignment with society's values when other players are still playing within a landscape of competing for finite resources and power. Think of the analogy: peaceful tribes who seek only to live in harmony with nature will just get killed by adjacent and competing warlike and extractive tribes. Without rules and guardrails, the most ruthless actors win. That's why legislation and policies are necessary, along with the collective will of the people to enact them. The greater meta-crisis is that these democratic guardrails-creating processes operate at a much slower pace than the rate of technological development that is needed to make a difference. Technology will only continue to advance faster than the harms can be well understood by twentieth-century democratic institutions. The technology sector itself needs to come together, collaboratively, and find ways to operate so that shared societal goals are placed above hyper-competition and profit maximization.

Finally, we need to recognize the massive asymmetric power that technology companies have over individuals and society. Any asymmetric power has to be in service of uplifting those beneath or with less power. That is the fiduciary model, or "duty of care" model. You simply cannot have an organization that has asymmetric power over you operating with a business model based on extracting profit from that relationship. Upgraded business models for technology need to be generative. They need to understand our incredible brilliance as well as our weaknesses, and support technologies that align with our deepest held values and humanity.

Towards Being Human

E. O. Wilson has said, "The problem with humanity is that we have Paleolithic emotions, medieval institutions, and godlike technology." We need to embrace our paleolithic emotions in all their fixed weaknesses and vulnerabilities. We need to upgrade our institutions to incorporate more wisdom, prudence, and love. And, we need to slow down the development of a godlike technology, with its powers that go beyond our capacity to steer the direction of the ship we are all on.

The realm of what is possible continues to expand, but it is co-arising with exponentially challenging global issues that require better information, leadership, and action. Rather than a race to the bottom that downgrades and divides us, we can co-create a technology landscape that enables a race to the top—one that supports our interconnection, civility, and deep brilliance. Change, I believe, is humanly possible.

WEALTH

The People's Bailout by Sam Wallman;
originally created for the Sunrise Movement

10

The Commons as a New Paradigm of Economics, Politics, and Culture

David Bollier

Let's remember 2020 as the year biological realities reasserted a more prominent, overriding role in the course of human affairs. The natural world warns us anew—with pointed urgency—that we are on the wrong track. It turns out that our audacious human inventions like the economy, state power, and technology are not autonomous machines that exist outside of history or the natural order. We are actually biological creatures, not just citizens or versions of *homo economicus* (the economic agent who rationalizes all choices toward economic gain). This is a shock to our consciousness because, as moderns, we do not readily acknowledge that we are profoundly interdependent on other organisms.

We thus face a new existential challenge: How can we make our modern, materialistic culture more compatible with a living, evolving planet? Despite our pretensions as champions of the Enlightenment, human life will not survive unless it moves more fully into sync with the ecological imperatives of the planet.

This insight requires us to probe more deeply into the nature of our political economy and culture. For mainstream players, the elephant in the room is neoliberal capitalism. By "capitalism," I don't just mean grand abstractions and macro-structures that seem quite divorced from our ordinary lives. The truth is quite the opposite: we are deeply implicated, personally, in the everyday practices and values fostered by capitalism. Yet there is no easy way to step outside of this pervasive system and its cultural categories and norms.

This is partly why the pandemic is so traumatic. COVID-19 is calling into question some basic conceits of modern life. Our faith in individualism over community, in humanity as separate from "nature," and in economic growth as "progress," are, quite simply, wrong. It's a bit superficial to say that the coronavirus is destroying the capitalist global economy. It's more accurate to say that it's destroying the epistemological edifice and cultural norms of "the economy" as we understand it.

If we are going to rebuild the world, we will need more than just new economic policies and governance institutions. We will need a new worldview, one that recognizes the world as a pulsating super-organism of living agents that happens to include humans. It will require that we shed the archaic foundational fictions of economics, for example, the fiction that defines human beings as self-interested utility-maximizers who are always seeking to increase material satisfaction.

In the biological world we inhabit, life is about striving for organic integration and wholeness. It is about deepening relationality and reciprocity. This is the deeper story of evolution, as it is explained by eco-philosophers like Andreas Weber. Highlighting these features of our reality is not simply an idle philosophical point. It's now a practical necessity. Present-day capitalism, which requires constant growth in order to avoid collapse, is going to destroy planetary ecosystems (and human civilization) within a few decades if it is not transformed or dismantled.

Fortunately, functional alternatives to capitalism already exist, most notably through *the commons*. But the social practices and ecological reasoning of *commoning* have little theoretical standing in respectable quarters, especially economics. Instead, our society's institutions continue to pay homage to the crumbling grand edifice of the market/state order, even though its structural deficiencies are becoming increasingly obvious.

The Deficiencies of the Market/State

COVID-19 and the 2008 financial crisis have drawn back the curtain on many myths used to justify the neoliberal capitalist narrative. It turns out that growth is not only impossible over the mid-term; it's also not widely or equitably shared. A rising tide does *not* raise all boats because the poor, the working class, and even the middle class do not benefit from the productivity gains, tax breaks, and equity appreciation that the wealthy enjoy. And yet taxpayers are commandeered to bail out banks and corporations that are "too big to fail" and to restore the supposed autonomy of "free markets."

The intensifying concentration of wealth is creating a new global plutocracy. Members of this elite group use their fortunes to dominate and corrupt democratic processes while insulating themselves from the ills afflicting everyone else. The "Invisible Hand" of the market was always a political fiction, but neoliberal policies of the past forty years have now made it a farce. This charade of commerce has systematically displaced costs and risks onto the poor, powerless social communities, ecosystems, and future generations. Inequality and precarity are expanding rapidly. Fear and resentment stoke white nationalism and authoritarianism under this economic strain. The market/state system and liberal democracy face a legitimacy crisis.

This is, of course, just a thumbnail account of a far more complex problem. But it's accurate enough to help us formulate a general critique. The growing inequities provide a compelling argument for a new sociopolitical imaginary that can go beyond what is on offer from the left or right. We need to imagine new sorts of post-capitalist governance and provisioning arrangements that can tame, transform, or replace predatory markets and capital accumulation and—in their place—support ecosystems, human care, social need, and personal development.

The regulatory state has failed to abate capitalism's relentless anti-ecological, anti-consumer, anti-social "externalities." Capital and financialization have eclipsed the power that the nation-state and citizen sovereignty once had to regulate the byproducts of market activity. The traditional left continues to believe, mistakenly, that warmed-over Keynesian economic policies, wealth redistribution, and social programs are politically achievable within existing the political regime—and likely to be effective. But in a democratic politics that is rigged and corrupted through campaign donations, social media disinformation, and problematic voting systems, it's getting harder to have faith in the efficacy of liberal representative democracy. These difficulties are compelling even apart from the realization that state bureaucracies and competitive markets are structurally *incapable* of addressing many societal problems. Their powers are too centralized, formal, and broad-gauged; they do not allow for the voices and initiatives of ordinary people. The limits of what "The System" can deliver on climate change, inequality, infrastructure, and democratic accountability have already been clearly demonstrated.

While the machinery of politics and government continues to function, its workings often amount to empty formalisms, symbolism, and staged propaganda. It resembles kabuki theater more than a robust vehicle for on-the-ground experimentation and collective action. It is fair to say that we now inhabit an "institutional void" of politics and policymaking. In the words of Dutch political scientist Maarten Hajer, "*There are no clear rules and norms according to which politics is to be conducted and policy measures are to be agreed upon.*" He continues: "To be more precise, there are no *generally accepted* rules and norms according to which policy making and politics are to be conducted."[1]

The Commons as a New Paradigm

The waning efficacy of the global market/state system has provoked countless activist and place-based communities around the world to develop a new path forward. These grassroots social movements take many forms: cooperatives, re-localization projects, food sovereignty, open source production, Transition Towns, degrowth, Social Solidarity Economy, care work, commons, and others. All of these movements emphasize different strategies with different styles, and yet the values of each overlap. All of them dissent from the grand neoliberal narrative of self-made individualism, expansive private property rights, economic growth, government deregulation, and consumerism.

While resistance to the neoliberal agenda is naturally a part of the shared agenda, these insurgent narratives—while diverse and sometimes fragmented—share a concern for building the actual alternatives themselves, rather than relying on politics and policy to deliver what's needed. System-change movements tend to share certain recurrent priorities:

- production and consumption for *use* as opposed to profit;
- bottom-up, decentralized decision making and social cooperation;
- stewardship of shared equity and predistribution of resources (i.e., by controlling equity assets themselves and not relying on state redistribution);
- an ethic of racial and gender inclusivism, transparency, and fairness;
- community, self-determination, and place-making over the purported imperatives of markets;
- a diversity of models adapted to local needs.

In these values we see a humanistic vision of society as a living, biodiverse system—not an inventory of resources to be allocated through

1. Hajer, "Policy without Polity," 175.

government agencies. The bottom-up movements stress the importance of *stewarding the earth* and its ecosystems; the *priority of people's basic needs* over market exchange and capital accumulation; and the importance of *participation, inclusion, and fairness* in successful resource management and community governance.

While there are many idioms for talking about these concerns, I have found that the language of the commons is particularly powerful both in critiquing neoliberal capitalism and in constructing a new approach to life. The commons offers a language for reorienting our perceptions and understanding. It helps name and illuminate the destructive realities of market enclosure and the enlivening value of commoning together. Without the discourse of the commons, these two social realities remain culturally invisible or marginalized—and therefore politically less consequential.

The language of commons provides a way to make moral and political claims that conventional policy discourse—certainly the major political parties—ignore or suppress. Using the concepts and logic of the commons helps bring into being a new cohort of commoners who recognize our mutual affinities despite our differences. Seed sharers and open source programmers, urban commoners and alternative currency designers, can all assert their shared values and priorities in systemic terms.

The coherent philosophical narrative of the commons can help prevent capital from playing one interest off against another. We no longer have to make false choices: nature vs. labor, labor vs. consumers, consumers vs. the community. Through the language and experiences of commoning, people in different social roles can begin to see their common identities. It offers a holistic vision that helps diverse victims of market abuse recognize their shared victimization, develop a new narrative, cultivate new links of solidarity, and build a constellation of working alternatives driven by a different logic and ethic.

Institutional Innovations to Support Commoning

The discourse of the commons goes a long way toward helping re-imagine the institutions of governance and provisioning in the post-pandemic world. Crucially, the commons gets us beyond stale debates about socialism versus capitalism. Both of these systems rely on problematic, centralized, hierarchical systems controlled by state power. The point of the commons is to open up new vistas for distributed action and thought that neutralize or bypass the capital/state alliance, rigid bureaucratic systems, and top-down, policy-driven approaches to change.

The commons looks to initiatives that are socially driven and that distribute risks and benefits in a mutual way without the costly overhead and hierarchies of the market/state. Localized governance empowers us and works to prevent the corruptions of consolidated power. Bottom-up, peer-driven organization and innovation give people a voice and engender trust. When collective energies and wisdom are mobilized around shared goals, social solidarity and stability are nourished.

The pandemic has revealed the fragility of global supply chains. Commons—more self-contained, place-based systems of provisioning—have shown how resilience can be achieved. One need only look at community supported agriculture (CSA), community land trusts, and local currencies to find examples of effective strategies for re-localizing value chains. Through an ecosystem of local or regional commons, it's possible to de-commodify our productive assets by removing them from the circuits of capitalist exchange. We can make them less dependent on volatile, expensive global markets (land, labor, technology), and we lessen dependence on outside finance by recirculating value locally (food provisioning, services, currencies, etc.).

CSAs are a time-proven finance technique for upfront sharing of the risk between users and producers. We know this as an agricultural finance tool, but in fact it can be utilized in many other contexts as well. In my region, many jazz fans subscribe to a series of jazz performances by paying upfront fees, CSA-style. Community land trusts (CLTs) are also a great way to de-commodify land. They take land off speculative markets permanently and mutualize the control and benefits of real estate. CLTs keep land under local control and use it for socially necessary purposes (for example, growing organic food locally) rather than for purposes favored by outside investors and markets.

The Schumacher Center for a New Economics has developed the notion of "Community Supported Industry," a strategy that attempts to substitute local production for products imported through global or national markets. The local community helps local businesses flourish by providing land through CLTs, local investment, worker training and education, and many other forms of support.

Another way to foster re-localization is through "Convert-to-Commons" strategies. These are novel financial or policy mechanisms that help convert private assets used for making profits into assets for collective use. An example is the British law that gives communities the first right to bid on community assets like a beloved pub or building for purchase. Another strategy, known as "Exit to Community," gives startup businesses an alternative to selling out to private investors or nasty big companies. Instead, they offer users or local communities the chance to purchase them.

Open source software communities have demonstrated new modes of production that are readily applicable to other realms. The idea is that national or global design communities can freely share and expand "light" knowledge—through open-source networks—while encouraging people to build the "heavy" (physical) stuff locally. This is sometimes called "cosmo-local production."

There are already a number of exciting examples of cosmo-local production arising for motor vehicles (Wikispeed car), furniture (Open Office), houses (WikiHouse), agricultural equipment (Farm Hack; Open Source Ecology), electronics (Arduino), and much more. Public Lab is a citizen-science project that helps address environmental problems by providing open source hardware and software tools, such as monitoring kits. Although at present they are often fledgling systems, cosmo-local forms of production have enormous potential to minimize the carbon footprint of conventional production while reducing transportation and intellectual property costs.

A related form of innovation is *platform co-operatives*. Internet platforms don't need to extract money from a community the way companies like Uber and Airbnb do. Rather, they can be vehicles for empowering workers and consumers. They can spur group creativity, reduce prices, and improve quality of life when they are designed and owned with the values of the commons in mind. Platform co-ops can improve operations and distribute market surpluses for the mutual benefit of participant-owners, instead of for absentee investors. We see examples of platform co-ops being used by taxi drivers in Austin, Texas (ATX Coop Taxi), food delivery workers in Berlin (Kolymar-2), delivery and messaging workers in Barcelona (Mensakas), and freelance workers in Brussels (SMart).

It's important that commoners can easily act as commoners without having to shoulder impossibly difficult or heroic burdens. That's why we need new forms of commons-based infrastructure to readily enable commons-based solutions. Infrastructure—physical, legal, administrative—provides a framework that makes it easier for individual commoners to cooperate and share. One example is Guifi.net, a commons WiFi system in Catalonia, Spain that has more than thirty thousand Internet nodes. Guifi.net provides high-quality, affordable service that avoids the predatory prices and business practices of corporate broadband and WiFi systems. Another interesting infrastructure project is the Omni Commons, an Oakland building that provides offices, studios, and meeting spaces for artisans, hackers, social entrepreneurs, and activists.

Beyond infrastructure, commoners will eventually need to rethink how law and policy can support commoning. This is a complicated topic because state power is generally more interested in promoting businesses and economic growth (to boost taxes and create jobs), even though commons can make a region more resilient, stable, and able to meet everyone's needs. Nevertheless, there is currently a great deal of interest in many cities—Amsterdam, Barcelona, Bologna, Seoul, San Francisco, among others—in creating urban commons as a counterpoint to top-down investment and speculation. Among other examples, rich fields of experimentation have recently emerged in the creation and management of public spaces, urban agriculture, public transport, social services, public events, and more.

We also need to reimagine conventional finance to support the distinctive work of commons. We need community-supported pools of money that are not encumbered with the growth imperatives that conventional interest-bearing debt and shareholder equity require. Fortunately, there are already many hardy examples to build upon: mutual aid societies and insurance, crowd-gifting and crowd-equity, financial models developed by community land trusts, CSA farming, and cooperative finance.

Commoning as a Way Forward

The strength of the commons lies in its core principles, porous boundaries, and endless permutations of possibility. These are core elements of any living system—especially social movements! And that's what makes the commons so valuable in dealing with COVID-19. We can begin to prioritize our biological, ecological, and creaturely needs. We don't need to remain hostage to a totalistic market/state system that is proving increasingly ill-suited for our lives and times. We can in fact leverage vernacular practice and culture to meet our needs in new ways. We can pursue serious transformations in provisioning, governance, and culture. We can re-imagine the configuration of state power and redefine the scope of economics to include our full humanity and social relations. Commoning nourishes people's instinctive needs for human connection and meaning, something that neither the current state or market can do.

The commons paradigm represents a deep philosophical critique of neoliberal economics and modern life, offering hundreds of functioning examples on which theorists and practitioners can build. Still, as an action-oriented approach to system change, it does not move forward automatically. Everything depends upon the ongoing energy and imagination that actual and would-be commoners contribute. Everything depends upon developing better infrastructures, legal regimes, and financial systems to facilitate the growth of commoning.

The anonymous Invisible Committee in France has observed that "an insurrection is not like a plague or forest fire—a linear process which spreads from place to place after an initial spark. It takes the shape of music, whose focal points, though dispersed in time and space, succeed in imposing the rhythms of their own vibrations." As a living system, commoning speaks to the heart in ways that ideologies do not. Its rhythms are producing a lot of vibrations. The question for the future is what types of new living institutions and social forms the resonance of commoning will catalyze.

References

Hajer, Maarten. "Policy without Polity? Policy Analysis and the Institutional Void." *Policy Sciences* 36.2 (2003) 175–95.

11

Public Banking as a Cornerstone
of the Commons

ELLEN BROWN

WE LEARNED A KEY lesson during the COVID-19 economic shutdown: the central bank can supply unlimited funds in a crisis to prop up the financial economy and the banks. At the peak of the pandemic, the Federal Reserve was making credit available to banks virtually interest-free. But those accommodations have not been extended to state and local economies. We the people have been bailing out the private banking system at least since the 1930s, when Congress agreed to underwrite the insolvent banks with deposit insurance from the Federal Deposit Insurance Corporation (FDIC). The FDIC shifted liquidity risks to the federal government and the taxpayers, but the American people—those footing the bill—are not sharing in the profits and the benefits. How can local governments stimulate their local economies and share in the favors showered on banks? I will argue that the answer lies in forming their own publicly-owned banks.

Tapping the Magic Money Tree—for Wall Street

Politicians have long argued that Medicare for all, a universal basic income, student debt relief, and a slew of other much-needed public programs are off the table because the federal government simply cannot afford them. But that was before Wall Street and the stock market were driven onto life-support by a virus. Congress has now suddenly discovered the magic money tree. Even Minneapolis Fed President Kashkari acknowledged that that the Federal Reserve has what amounts to an infinite amount of cash, and Congress unanimously passed the Coronavirus Aid, Relief, and Economic Security (CARES) Act in only a few days. The bill authorized *$2.2 trillion* in crisis relief, most of it going to Corporate America with very few strings attached. The Federal Reserve also opened the floodgates to unlimited quantitative easing (QE), stating that it would buy Treasury securities and mortgage-backed securities "in the amounts needed to support smooth market functions."

The extraordinary expenditures authorized by Congress and the central bank were allegedly necessitated by the COVID-19 crisis and intended to relieve the people and the economy of the devastating effects of a forced shutdown. But little of this money has actually gone to individuals, families, communities, or state and local governments. Qualifying individuals got a very modest one-time payment of $1,200, and unemployment benefits were extended for several months. For local governments, $150 billion was allocated for crisis relief, and one of the Fed's newly expanded Special Purpose Vehicles was set up to buy municipal bonds. But the interest charged is at a penalty rate and the terms are restrictive, which means few states are likely to take advantage of this facility.

Rather than saving the real, productive economy, most of the bailout money has gone to saving the "financialized" economy, where the financial crisis actually hit long before the virus did. Banks largely stopped lending to each other in the Fed funds market after 2008 because they no longer trusted the borrowing institutions to have the money to pay up, and because the Fed started paying interest on the banks' excess reserves, giving them little incentive to go elsewhere for income on their idle funds. Banks turned instead to the repo market for the liquidity they needed to satisfy customer withdrawals. But lenders largely quit lending in the repo market in September 2019, when repo interest rates shot up to ten percent. To save the banks, the Fed felt compelled to step in to backstop that market. By March, it was making $1 trillion a day available for repo loans, an untenable situation.

Why is bank access to "liquidity" so critical to the economy that trillions of dollars must be conjured up in a crisis just to keep the financial system afloat? The answer underscores a fundamental flaw in the private banking model: banks don't actually have the money they lend. They simply create it by writing numbers into the accounts of their borrowers, as the Bank of England confirmed in its first 2014 quarterly report. Banks claim to be lending money received from deposits, but they also promise the depositors that their money is available on demand. When depositors and borrowers come for their money at the same time, the banks must borrow from somewhere else.

In March 2020, the Fed solved the liquidity problem by making credit available at its discount window to all banks in good standing at 0 percent to 0.25 percent—virtually free—with no penalty and no strings attached. Gone is the pretense that banks are drawing from a fixed, private pool of funds to be used as they will. They are drawing from a very public pool of liquidity created on the books of the central bank, backed by nothing but the full faith and credit of the United States and its people. It's our money, but it has been captured by a private banking cartel that is now using it for the banks' own purposes. For a sustainable solution to our economic crises, banking needs to be made a public utility.

Time for the States to Step In:
The Model of the Bank of North Dakota

State and local governments today are second-class citizens when it comes to borrowing. Unlike the banks, which can borrow virtually interest-free with no strings attached, states can sell their bonds to the Fed only at penalty rates. Yet states are excellent credit risks—far better than banks would be without the life-support of the federal government. States have a tax base; they are legally required to pay their bills; and they are forbidden to file for bankruptcy. Banks are considered better credit risks than states only because their deposits are insured by the federal government and they are gifted with routine bailouts from the Fed. Without these publicly backed benefits, they would have collapsed decades ago.

State and local governments, which have a mandate to serve the public, deserve to be treated at least as well as private Wall Street banks, which have been found guilty of frauds against the public. If Congress won't address this disparity, state and local governments need to do it themselves. They can borrow interest-free from the Fed by forming their own publicly-owned banks.

The stellar model in the United States is the Bank of North Dakota (BND)—currently our only state-owned bank. Founded in 1919, the BND has served its state brilliantly. In November 2014, *The Wall Street Journal* reported that the BND was more profitable than even the largest Wall Street banks, with a return on equity that was seventy percent greater than that of either JPMorgan Chase or Goldman Sachs.[1]

Ironically, the goal of the BND is not actually to make a profit. It was formed in 1919 to free farmers and small businesses from the clutches of out-of-state bankers and railroad men. Its stated mission is to deliver sound financial services that promote agriculture, commerce, and industry in North Dakota.

Why then is it so profitable? Its secret is its efficient business model. Its costs are low: no exorbitantly paid executives; no bonuses, fees, or commissions; no private shareholders; low borrowing costs; no need for multiple branch offices; and no FDIC insurance premiums (the state itself, rather than the FDIC, guarantees its deposits). BND profits are not siphoned off to Wall Street or stored in offshore tax havens. Instead, profits are recycled back into the bank, the state, and the community.

All of the state's revenues are deposited in the bank by law, giving it a massive captive deposit base. Most state agencies must also deposit with the BND. It does not compete with local North Dakota banks for the deposits of individuals or municipal governments. Instead, it helps local community banks by providing letters of credit that guarantee municipal government deposits in local banks.

The BND also partners with local North Dakota banks rather than competing with them for loans. The local bank acts as the front office dealing directly with customers. The BND acts more like a "bankers' bank," helping with liquidity and capital requirements. Local banks are thus able to take on projects that would otherwise either go to Wall Street banks or go unfunded.

The BND acts as a mini-Fed for the state. It provides correspondent banking services to virtually every financial institution in North Dakota. Because it assists local banks with mortgages and guarantees their loans, local North Dakota banks can keep loans on their books rather than offload them to investors to meet capital requirements. This helped local banks avoid the subprime and securitization debacles. The state has reduced the need for wasteful rainy-day funds invested at minimal interest in out-of-state banks through a cheap and ready credit line with the BND. When North Dakota went over-budget in 2001 due to the dot-com crisis, the BND acted as a rainy-day fund for the state. And when a local city suffered a massive flood in 1997, the bank provided emergency credit lines.

Due to this amicable relationship, the North Dakota Bankers' Association endorses the BND as a partner rather than a competitor of the state's

---

1. Dawson, "Shale Boom."

private banks. Indeed, it may be the BND that ultimately saves local North Dakota banks from extinction as the number of banks in the United States steadily shrinks.

In the COVID-19 economic shutdown, North Dakota's effective banking system was on full display. It has topped the list in gaining access to federal funds for individuals and local businesses, largely due to its strong community banking system backed by its state-owned bank. Coordinated by the BND, North Dakota's community banks distributed unemployment benefits ten times faster than the slowest state, and small businesses in North Dakota secured more Payroll Protection Program funds per worker than in any other state. Other COVID-19 relief from the BND includes deferment of payment on student loans for up to six months, small uncollateralized business loans up to $25,000 at a fixed interest rate of 1 percent, and larger business loans up to $10 million at 3.5 percent with the option to reduce the rate to 1 percent.

Like public sector banks generally, the BND lends counter-cyclically. After the 2008 banking crisis, while Wall Street banks were being bailed out by the taxpayers and were drastically cutting back on local loans, the BND was increasing its local lending—and at the same time showing record profits. In the last eighteen years, it has reported an average annual return on equity of 20.2 percent.

The BND passes its profits on to North Dakotans, both as dividends to the state and by facilitating below-market loans. Students can get favorable loans from the BND that come without fees or the onerous conditions often imposed by private lenders. In 2015, the North Dakota legislature established a BND Infrastructure Loan Fund program that made $50 million in funds available to communities with a population of less than two thousand, and $100 million available to communities with a population greater than two thousand. The loans were to have a two percent fixed interest rate and a term of up to thirty years. At that time, the taxable rate on infrastructure bonds in other states was four to six percent. The loans could be used for the construction of new water and treatment plants, sewer and water lines, transportation infrastructure, and other infrastructure needs to support new growth in a community.

Other Public Banking Models

The US has only one state-owned bank, but over twenty-one percent of banks globally are publicly owned. In the 1970s, before the push for privatization and "liberalization" of banking laws, more than fifty percent of bank assets worldwide were controlled by the public sector.

Public sector banks are particularly successful in Germany, where they dominate the local banking scene. Like the BND, these local savings banks, or *Sparkassen*, actually outperformed the private banking sector while serving the public interest. The *Sparkassen* operate a network of over 15,600 branches and offices, employ over 250,000 people, and have a strong record of investing wisely in local businesses. The *Sparkassen* network capitalizes on economies of scale to provide services to its members, including a compliance department that deals with the onerous regulations imposed on banks by the EU.

Germany has been called "the world's first major renewable energy economy," and public sector banks provided most of the financing for this energy revolution. During the COVID-19 financial crisis, German public development bank KfW has been making business loans available through a simple application process at a very modest one percent interest.

Another public bank that is popular worldwide is the postal savings bank. According to the Universal Postal Union, 1 billion people now use the postal sector for savings and deposit accounts, and more than 1.5 billion people take advantage of basic transactional services through the post. Since maintaining post offices in some rural or low-income areas can be a losing proposition, expanding their postal business to include financial services has been crucial in many countries to maintaining the profitability of their postal networks. Public postal banks are profitable because their market is large and their costs are low. The infrastructure is already built and available, advertising costs are minimal, and government-owned banks do not reward their management with extravagant bonuses or commissions that drain profits away. Instead, profits return to the government and the public. Postal banking systems are a fixture in other countries, where their long record of safe and profitable banking has proven the viability of the model.

A century ago, postal banking was also popular in the United States. In the late nineteenth century, a nationwide coalition of workers and farmers united to demand postal savings banks of the kind found in most other nations. They argued postal banks could provide depositors with basic banking services and were a safe haven against repeated financial panics and bank failures. After the private banking system crashed the economy in the Bank Panic of 1907, a postal savings alternative was established by the Postal Savings Bank Act of 1910. The US Postal Savings System was set up to get money out of hiding, attract the savings of immigrants, provide safe depositories for people who had lost confidence in private banks, and furnish depositories with longer hours that were convenient for working people. The postal system paid two percent interest on deposits annually. It issued US Postal Savings Bonds that paid annual interest, as well as Postal Savings Certificates and domestic money orders. Postal savings peaked in 1947 at almost $3.4 billion.

The US Postal Savings System came into its own during the banking crisis of the early 1930s, when it became the national alternative to a private banking system people could not trust. But postal banking was under continual assault from the private banking establishment. When demands increased to expand its services to include affordable loans, alarmed bankers called it the "postal savings menace" and warned it could result in the destruction of the entire private banking system.

Rather than expand the Postal Savings System, President Franklin Roosevelt responded by buttressing the private banking system with public guarantees, including FDIC deposit insurance. Private banks were now in the enviable position of being able to keep their profits while their losses were covered by the government. Deposit insurance, along with a statutory cap on the interest paid on postal savings, caused postal banking to lose its edge. In 1957, the head of the government bureau responsible for the Postal Savings System called for its abolition, arguing that "it is desirable that the government withdraw from competitive private business at every point." The Postal Savings System was finally liquidated in 1966.

The push for privatization of the US Postal Service continues today. Although the USPS has been successfully self-funded without taxpayer support throughout its long history, it is currently struggling to stay afloat. What has driven it toward insolvency is an oppressive congressional mandate, included almost as a footnote in the Postal Accountability and Enhancement Act of 2006 (PAEA), which requires the USPS to pre-fund healthcare and pensions for its workers seventy-five years into the future. No other entity, public or private, has the burden of funding multiple generations of employees yet unborn. The pre-funding mandate is so blatantly unreasonable that the nation's largest publicly-owned industry appears to have been intentionally targeted for takedown.

The USPS could recapitalize itself by reviving the banking services it efficiently performed in the past. A January 2014 white paper published by the Office of the Inspector General of the USPS argued that a system of postal banks could service the massive market of the unbanked and underbanked, which includes about one in every four households. Without access to conventional financial services, people turn to an expensive alternative banking market of bill-pay, prepaid debit cards, check cashing services, and payday loans. They pay excessive fees and are susceptible to high-cost predatory lenders. Catering to this underserved group would not only be a revenue generator for the post office but would also save the underbanked large sums in fees.

In April 2018, Sen. Kirsten Gillibrand introduced legislation that would require every US post office to provide basic banking services. In May 2019, Sen. Bernie Sanders and Rep. Alexandria Ocasio-Cortez brought the Loan Shark Prevention Act, which would cap the interest charged by credit cards and payday lenders at fifteen percent and would turn post offices across the nation into providers of low-cost basic financial services, including checking accounts, savings accounts, and some loans. These are measures we can take immediately to address the public service gaps in our banking and financial industries.

Money Creation as a Tool for Public Good

COVID-19 is shining a bright light on the inequities, fragilities, and flaws in the current banking scheme. The recent financial crisis has generated renewed interest in the expansive possibilities of a financial model freed from the shackles of neoliberal monetary theory and policy. If we can come up with trillions of dollars to save the financial economy, as the United States has, we can come up with the money to save the real, productive economy. The real economy desperately needs a massive injection of new money— something that a publicly-owned banking system can provide that while also serving the public good.

A movement for publicly-owned banks is gaining momentum in the US, with over twenty-five bills to set up or study them now being pursued, and over fifty groups advocating for them across the country. For more information, see PublicBankingInstitute.org.

References

Dawson, Chester. "Shale Boom Helps North Dakota Bank Earn Returns Goldman Would Envy." *The Wall Street Journal*, November 16, 2014. https://www.wsj.com/articles/shale-boom-helps-north-dakota-bank-earn-returns-goldman-would-envy-1416180862.

12

Building an Economy
of Wellbeing and Indigenomics

MARK ANIELSKI

THE CHINESE SYMBOL FOR crisis has two meanings: danger and opportunity. According to Professor Victor Mair, an expert in Chinese language and literature at the University of Pennsylvania, the Chinese ideogram for crisis is *wēijī*, which means something like "incipient moment; crucial point (when something begins or changes)." Mair points out that "a *wēijī* is indeed a genuine crisis, a dangerous moment, a time when things start to go awry. A *wēijī* indicates "a perilous situation when one should be especially wary."

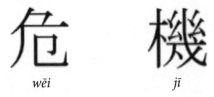

wēi　　　　　*jī*

The world stands at the tipping point of systemic crisis: health (COVID-19), financial, institutional, climate, and ethical. I want to explore the possibility of adopting a new economic paradigm from the perspective of an ecological civilization. I would like us to think about what it means to live together in an economy of wellbeing (what I call Genuine Wealth) and ultimately love.

An Ecological Civilization: A New Economy

Dongwoo Lee (the Director of EcoCiv Korea), provides an elegant description of an ecological civilization that has found considerable attention in China and Asia. His article suggests that the current neoliberal economic system is fragmenting and showing its fragility.[1] In response to this situation, Lee asks: What does an economic system after COVID-19 look like?

Like Lee, I believe we are in the midst of perfect storm to consider a new sustainable system. We can form economics around ecological principles, natural laws, and the idea that wellbeing, compassion, and love are common human aspirations for an economy. After all, the word economy (*oikonomia* in Greek) refers to the wise management (*nomia*) of the household (*oikos*).

In my books *The Economics of Happiness: Building Genuine Wealth* and *An Economy of Wellbeing: Common Sense Tools for Building Genuine Wealth and Happiness*, I presented a new economic framework for measuring progress based on wellbeing rather than economic growth measured by gross domestic product (GDP). I believe that an economy of wellbeing is possible based on mimicking nature and grounded on Indigenous natural laws.

The Global Debt Balloon Catastrophe Looms

A growth economy is necessary to fund a debt-based money system that demands increasing interest charges on an exponential mountain of debts. When we examine the growth of total outstanding debt (household, government, and business) we see and explosion since 1946. Figure 1 shows the total debt outstanding in the US economy which has now grown an incredible 8,000 percent from 1946 to the first quarter of 2020 reaching $77 trillion, according to US Federal Reserve Statistics. When you compare annual GDP with total debt, there is a strong correlation that shows debts are never repaid and GDP must continue to grow. This comes at the cost of the destruction human wellbeing and the environment.

In a growth economy, money is debt. Debt money imposes a growth bias in the world's global economic model. Few economists have ever examined the relationship between debt money and GDP. In 1988, I was encouraged to examine this relationship by Herman Daly. He suggested that understanding this relationship was the most important, unresearched inquiry in economics.

1. Lee, "The Next Economy."

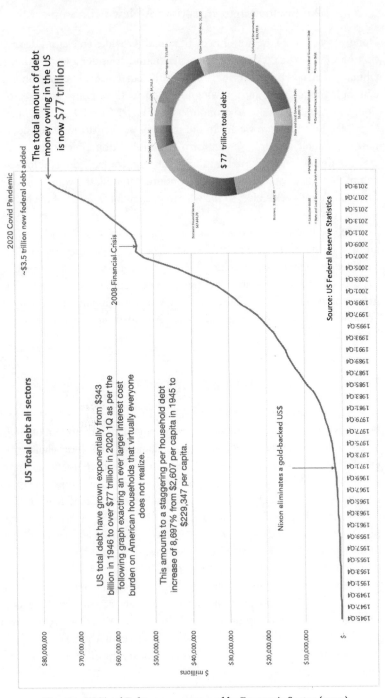

Figure 1: US Total Debts 1945–2020 and by Economic Sector (2020)

To do this, we only need to look at the ratio trends of total debt to GDP for the US and other nations. In the US, this ratio has grown from an average and stable 146 percent (total debt to GDP) between 1946 and 1971 (the year Nixon removed the US dollar from the gold standard) to 362 percent as of the end of 2019. Based on the most current IMF (International Monetary Fund) data, other countries have similar total debt (government and private/household) to GDP ratios as the US: Canada: 356 percent, Luxembourg: 496 percent, Japan: 445 percent, Ireland: 414 percent, Portugal: 368 percent, South Korea: 358 percent, France: 351 percent, the UK: 310.8 percent, and Germany: 216 percent.

Conventional economic wisdom suggests that once national debt (government debt) exceeds 100 percent, it becomes a cause for concern. But what do total debt to GDP ratios (household, business, and government) that exceed 300 percent suggest? Is it sustainable for an economy? We have entered unprecedented economic turbulence; this a state of dead-man-walking or zombie economics.

It's important to understand that there are no apparent solutions for taking the air out of an ever-inflated debt balloon. It also appears that there is no recognition that these levels represent a potential near-death experience for all world economies. Strikingly, these debts are unrepayable given current economic output/GDP. Before COVID-19, GDP had to grow just to service the ever-increasing interest payments on exponential debts. This means economic consumption and GDP growth could never stop. If it did, the entire debt system would suffer a catastrophic heart-attack like we saw after the 2008 financial crisis (seen in Figure 1 as the US debt curve went horizontal or flat-lined for the first time since 1946). I've calculated the growing cost of interest payments on total debts in the US and Canadian economies (something rarely considered by economists) on total debts which have been doubling every seven to eight years in the US since 1946.

The debt costs associated with paying for the economic fallout from COVID-19 are expected to be huge. It will push the US, Canada, and other nations to even higher unprecedented levels of debts that can never be repaid from future GDP growth. Like a malignant cancer cell, debt will ultimately consume the host.

The debt cancer affects every other nation on Earth—with perhaps the exception of China. According to Ellen Brown, a US expert in public banking and monetary policy, China has adopted the financial sovereignty model developed by Abraham Lincoln at the dawn of the twentieth century. If so, China may be the only nation to have a truly sovereign wealth system that may withstand the impending debt implosion.

This debt cancer is the ultimate threat to the neoliberal economy that is based on perpetual economic growth, unsustainable trade, climate catastrophe, and debt-money systems that perpetuate severe income and power inequalities.

Hidden Costs of Debt to Average Households

The burden of these interest payments on the average American household currently consumes a staggering $47,674 or 75.6 percent of median US household income. While there are no official interest cost estimates reported by governments or national statistical agencies, these statistics remain hidden and unnoticed. These estimates suggest the average American household is working nearly thirty hours of a typical forty-hour work week just to pay for these hidden interest charges on all debts outstanding for the American economy.

Measuring What Matters to Wellbeing?

Some progressive economists since the mid-1970s have developed alternative measures of progress. These include the Genuine Progress Indicator (GPI) that attempts to distinguish economic activities that contribute to wellbeing from those that detract. This full-cost accounting approach deducts societal costs of income and wealth inequality, crime, time contributions for raising children or caring for elders, and environmental degradation from GDP. Alternative measures of progress include the Canadian Index of Wellbeing, Bhutan's Gross National Happiness, and the OECD Better Life Index used by European nations and New Zealand in their approach to wellbeing-based budgeting.

From a standard accounting perspective, nations lack a fulsome balance sheet to measure wellbeing conditions. To rectify this shortcoming, I have proposed the development of a more comprehensive system called Genuine Wealth. Genuine Wealth accounts for the physical, qualitative, and monetary conditions of wellbeing for the five key assets of any nation, province/state, municipality or corporation: human, social-cultural, natural, built, and financial-economic assets.

Measures of wellbeing include objective statistics and subjective indicators from annual surveys to assess societal wellbeing of citizens. I have developed new wellbeing survey tools that allows citizens to self-evaluate their mental, physical, emotional, spiritual, and economic wellbeing. Subjective measures are aligned with objective statistics from national statics agencies which provide a balanced account of the lived and experiential conditions of the citizenry's wellbeing.

We can also maintain a set of accounts for the qualitative state of the natural environment (on the scale of a watershed or ecoregion). Natural capital accounting has advanced since the early 1990s and most nations now have some preliminary can assess the health of ecosystems. This includes measures of the vitality of ecosystem goods and services. Here again, Indigenous peoples can serve a key role in assessing and verifying the health of forests, wetlands, rivers, and other ecosystems based on their traditional knowledge and understanding of the health of these natural systems.

When we account for the wellbeing of a nation's assets, wiser decisions are possible as we plan and budget so true capital and operating expenditures can be quantified. This provides the basis of determining the value of programs and services. By measuring impacts on wellbeing for a community or nation, measures of progress can extend beyond monetary assessments like GDP or business profits.

What Should We Use as Measures of Progress?

As Dongwoo Lee reminds us, the belief (or mythology) in traditional economics is the goal of a business is to make profit and the purpose of an economy is to generate positive GDP growth. But if we examine the origins of accounting in sixteenth-century Venice or revisit Aristotle's teachings on an economy, we discover that profit maximization has never been the basis of a business enterprise. In fact, GDP maximization and economic efficiency was considered a short-term necessity in measuring economic reconstruction after World War II by the key architects of national accounting of that time (John Maynard Keynes, Simon Kuznets, and others). Kuznets suggested that measuring money exchange and consumption in an economy would eventually be overshadowed by a focus on quality of life, nutrition, health, and resilient ecosystems.

Economists and accountants see assets as ownership of value that can be converted into cash (although cash itself is also considered an asset). We can broaden the definition of an asset to include anything tangible or intangible that contributes to the wellbeing of an individual, an organization, the community, a nation, and natural ecosystems. This allows accountants and economist to expand the scope of measurement and reporting to including intangible assets. We can now include trust, goodwill, relationships, and other forms of social capital along with natural capital assets (forests, wetlands, carbon) and ecological goods and services. These wellbeing assets would then become the basis of a sustainable livelihoods (as per the Brundt-land Commission model of 1989) for individuals and communities.

Agreeing upon a common set of values across diverse cultures will be difficult. Values are unique to individuals, cultures and communities. A common accounting system for measuring wellbeing that is applicable to all nations or communities is hard to establish. However, expanding our understanding of assets, can provide a common platform for how nations and communities might look at themselves. It offers progress for how conditions of wellbeing are evaluated.

A New Genuine Wealth Model

Lee's vision of an ecological civilization recognizes that our world is in the midst of a serious pandemic, the world economy is paralyzed, and we continue without alternatives to the debt-money-cancer economies. Lee offers a new fundamental framework—a new economic system—in which we see money as a tool, not a purpose. His vision aligns with my framework of Genuine Wealth (Figure 2).

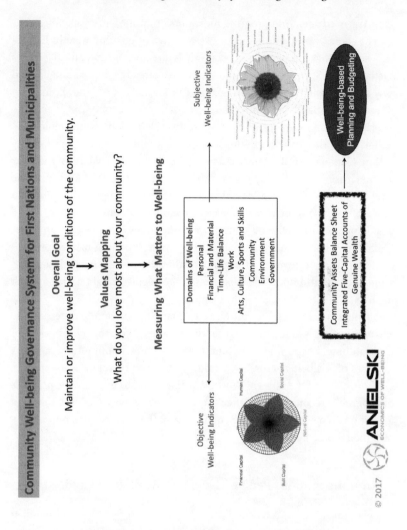

Figure 2: Genuine Wealth Model for Measuring Wellbeing

Genuine Wealth is based on the harmony of two words: genuine (to live authentically according to common values and virtues) and wealth (a thirteenth-century Old English word for "the conditions of wellbeing").

By orienting our conversation around original definitions of the words *economy* and *wealth* (i.e. wellbeing) there is a path forward for an ecological civilization. We have a practical working model of an economy of wellbeing. Lee notes (and I agree) that we need a new system of values along with new indicators of success and progress. Our new economic system must help humanity thrive without destroying the environment.

We need "genuine progress indicators" that measure wellbeing conditions across various domains or asset classes of genuine wealth. Lee acknowledges that the "wellbeing economy is already conceptualized." This is partially true. Some countries and cities have already adopted and applied new systems using wellbeing indicators (New Zealand, Iceland, Edmonton [Canada], Scotland, and now Amsterdam [donut economics]), but the implementation and execution of these models remain rather tepid. The wellbeing models have yet to impact monetary and banking policies of any nation. Canada like Bhutan has developed an Index of Wellbeing similar to the Gross National Happiness indicators of Bhutan; however, not one level of government, from federal to municipal, has adopted the Canadian Index of Wellbeing or other subjective measures of happiness that would be comparable to GDP and conventional measures of economic progress. Bhutan remains a quaint and small outlier amongst nations blessed perhaps by its spiritual uniformity of Buddhist values that form the philosophical foundations that support the happiness of the people in right harmony with nature as the ultimate goal of any economy.

There continues to be resistance to changing the dominant economic and accounting systems. Notwithstanding, there is hope that new pockets of innovation and experiments in building economies, communities, and enterprises of wellbeing across the world will form a critical mass for change. GDP arose out of the ashes of World War II as a tool to measure the progress of economic reconstruction after the cataclysm of war. COVID-19 and a looming ecological catastrophe might yet be the catalyst for embracing a new model based on wellbeing.

An Indigenomics Perspective on an Ecological Civilization

Indigenomics is a word recently coined by Carol Anne Hilton, a Canadian Indigenous economist and business leader. Carol Anne is restoring an ancient understanding of Indigenous economies unique to Indigenous cultures around the world. Indigenomics sees all things as interrelated: plants, animals, people, and ecosystems.

Each person and the community as a whole are seen as sacred circles (a "medicine wheel"). Indigenomics sees all assets as shared amongst the members of a community. The vision of the potlatch—an annual ceremony of sharing in the abundance of individual family or clan material wealth with other families—reflects the abundance witnessed in nature. The potlatch is a kind of break on the incipient threat and anxiety of scarcity and a potential threat of greed. Contrast these images with that of the linear models of debt and GDP described above.

Indigenous laws—similar to natural laws defined in Western economies—guide decision making and determine what assets are vital to a good life (wellbeing). Natural laws, including the laws of water, were mostly upheld by the women of Indigenous communities. In addition to natural laws, spiritual laws formed the values and virtue foundations of these communities—though they were never codified.

Working with Indigenous communities in Canada, I envision restoring the original elaborate systems of economic exchange and governance that allowed the millions of Indigenous cultures on Turtle Island (North America) to thrive and flourish for ten thousand years. A remnant of these wisdom traditions still exists despite efforts to extinguish the "Indian" spirit and people. Building accounting and governance systems on a solid foundation of Indigenous values, principles, and laws is fundamental to good governance. It is congruent with the notion of a wellbeing-based economy and an ecological civilization.

Wealth (defined as wellbeing) is consistent with an indigenomics view of economies. From an Indigenous perspective wellbeing considers the mental, physical, emotional, and spiritual wellbeing of individuals, families and the community as a whole. From an indigenomics perspective, all wealth or wellbeing is considered to have originated from the Creator or God. In every ceremony and meeting an elder acknowledges that life is a gift of the Creator. Ownership or property rights by individuals is incompatible with the Indigenous view of shared responsibility. The Indigenous view emphasizes the stewardship of the total or genuine wealth of a nation or community, in harmony with Mother Earth or Nature.

All conditions of wellbeing and assets are seen as an integrated whole requiring a holistic framework for measuring progress. Measures of wealth from an indigenomics perspective includes a balance of quantitative and qualitative (or subjective) measures of wellbeing. Again, this points to the importance of accounting for wealth according to the original definition, namely measuring the wellbeing conditions of the assets that contributes to the long-term wellbeing of an enterprise, a community, or a nation. Ironically, virtually every government at any level (federal, provincial, or municipal), including First Nations, fail to produce a comprehensive asset balance sheet that would report the wellbeing conditions of genuine wealth of the community.

This more comprehensive definition of wealth and wise stewardship requires an integral/holistic framework that recognizes the relationality of all assets. All assets are interconnected, in relationship, and work together to cultivate a condition of flourishing. New visual images and presentations of wellbeing (data using circle graphs or diagrams) provide a useful way to represent these indicators and data. It helps showcase the interconnectivity and mutuality of a nation's assets and the relationship between indicators of wellbeing.

Conclusion

An economy of wellbeing is within our grasp. COVID-19 and its economic repercussions represents opens space to explore a better global economic system of indigenomics based on our yearning for a full and happy life. We must leave our debt straight-jacket behind. This requires a global dialogue about alternative money systems without debt, founded on the pragmatic ideas of an ecological civilization modeled after natural ecosystems. Natural ecosystems epitomize resilience, homeostasis, restoration, mutuality, and harmony. They repel monocultures and egoism and embrace diversity.

COVID-19 gives us an unfiltered look at our current economics and calls us to come up with a new system of values. We must realize that we are connected to each other and unite our powers and enliven new movements.

I would like to propose an even grander aspiration. I sense a deeper yearning for what I envision as a civilization of love. The word love is as complex as wellbeing, but I believe the fundamental principle of love is "caring for the wellbeing of another" (including the environment). A civilization of love is founded on a common human characteristic of compassion, empathy, and altruism. Of course, this runs counter to the neoliberal values of capitalism that see humans as egocentric, hedonistic, and materialistic. The ultimate building block of any economy and civilization is love.

I want to share a vision that may leave us with some hope for what is possible. The vision is about a Sacred Tree and comes from a gathering of several Indigenous elders in Alberta, Canada in 1981:

> For all the people of the Earth, the Creator has planted a Sacred Tree under which they may gather, and there find healing, power, wisdom and security.

> The roots of this tree spread deep into the body of Mother Earth. Its branches reach upward like hands praying to Father Sky.

> The fruits of the tree are the good things the Creator has given to the people; teachings that show the path to love, compassion, generosity, patience, wisdom, justice, courage, respect, humility, and other wonderful gifts.

The ancient ones taught us that the life of the Tree is the life of the people.

If the people wander far away from the protective shadow of the Tree, if they forget to seek the nourishment of its fruit, or if they should turn against the Tree and attempt to destroy it, great sorrow will fall upon the people.

Many will become sick at heart.
The people will lose their power.
They will cease to dream dreams and see visions.
They will begin to quarrel among themselves over worthless trifles.
They will become unable to tell the truths and to deal with each other honestly.
They will forget how to survive in their own land.
Their lives will become filled with anger and gloom.
Little by little they will poison themselves and all they touch.

It was foretold that these things would come to pass, but that the Tree would never die.
And as long as the Tree lives, the people live.

It was foretold that the day would come when the people would awaken, as if from a long, drugged sleep; that they would begin, timidly at first but then with great urgency, to search again for the Sacred Tree.

The knowledge of its whereabouts, and of the fruits that adorn its branches have always been carefully guarded and preserved within the minds and hearts of our wise elders and leaders.

These humble, loving and dedicated souls will guide anyone who is honestly and sincerely seeking along the path leading to the protecting shadow of the Sacred Tree.

Reference

Lee, Dongwoo. "The Next Economy: Transforming Economics Systems after COVID-19." Rev. Dongwoo Lee. Accessed August 12, 2020. http://www.revdongwoo.com/2020/04/30/코비드19-이후의-글로벌-경제-체계의-변화-the-next-economytransforming-economics-systems.

WORK

The Hive by **Abby Paffrath**

13

The Means Are the Ends
Microeconomics Will Matter

JESS RIMINGTON

Did you hear the story of the woman who gave birth on the water buffalo?

It was during a typhoon in the Philippines, and the ferocity of the storm induced her labor. She was in the flooding waters. A water buffalo came by and she grabbed hold of her horns. The buffalo helped the woman get up on her back and swam onwards. As they were swimming, they passed another woman on top of a piece of a car. From across the new river the two women talked with one another. They discovered one was in labor and that the other happened to be a midwife. The midwife jumped on the water buffalo and guided the woman in labor as she gave birth.

. . . It's something made for folklore: *the world-that-was, shattered.* The thousand small moments of serendipity to summon the midwifing of this one birth . . . improbable? I can't confirm or deny this series of events.[1] But I've been told that the baby is a girl, that she is named after the typhoon, and that she smiles a lot.

---

1. This is a story that was shared with me by a midwife about another midwife.

∾

Sometimes, these days, I meet the water buffalo in my dreams. She lowers her head to lend a corrugated horn. I lay on her back facing the purple night sky, stars winking blue-white, and we speak of times that could be: if we were to grab onto each other, if we were to—

    be midwives of transition,

    to different ways.

She's told me in the "otherwheres"—the times when we've done the hard things: the reckoning, the repair, the birthing—that I smile a lot too.[2]

∾

I've been thinking a lot about death, probability, and imagination. Death, because of all that we've endured this year (2020). Because I'm committed to keeping the passage from this life sacred. Committed to death never becoming just a statistic.

Probability. . . because it's now part of the constant micro risk-assessment calculations I make while doing once-rote tasks, like grocery shopping.

And imagination? . . . Because I find my body asking, "how do we *un*imagine these ways?" The ways that built the unjust systems that, at best, continue to fail us and, at worst, oppress and murder. The pandemic has further laid bare the deep fault lines and cruelties of our current economy.

When I close my eyes, I see the threatening, invasive briar of our dominant paradigm stretching out to cover land after land. Faster and faster it goes, as if afraid it won't have time to absorb and devour everything. There's incentive to capture, entwine, and then render our diverse spaces of human activity into flatness—to trap the symphony of what is possible in its barbs . . . and then smash it with soundbites, overly long hours of work, fear, and scarcity, until it's level, smooth, implacable, and quiet. Dead. To somehow bleed us of the innateness of other *possibles* by using mythologies that separate us from each other. To leave us with little energy to be alive.

As it turns out, the resulting homogeneity is quite good for commerce, and even better for the extraction of labor and resources. The flatter or more adamantly "settled" things seem, the more it can feel like particular ideological choices are *not* actually so choiceful. "This is the way things are," we are told, as though our current paradigm is a natural state of being.

---

2. Henceforward, "Meditating."

This inequitably violent and extractive system is at once real and ensnaring, yet also as fragile as any ideology.

Seeing "economy" as a concept available to us apart from a particular "-ism" can be uncomfortable. *What is an economy outside of capitalism, socialism . . . or even any -ism?* If we can even get our brain to go there, it can feel like calling out "economy" as potent for more diverse renderings incriminates us. "Sssssh!" Sedition. Keep them preoccupied with survival, so they don't look. *The emperor might be naked.*

As radical as it may sound, economies are products of our imagination. They are social constructs. We make them real by behaving according to our belief that they exist, and thus actualizing them. This means we re-create the economy every day. We re-*create, everyday,* the white supremacy and patriarchy encoded within our economic systems, the inhumane treatment in our workplaces, the constant time crunch, the so-called "externalities" of lost human life and environmental degradation. We re-*create* the hamster wheel. The economic rules to which we have collectively assented are imagined—even though their impacts are undeniably real. This is especially true for peoples and communities who have long seen behind the curtain and fought back, some for their very survival.

There's a phrase for the shared space many minds conjure together to exist within—the kind of space that dictates our templates for how we behave together. It's what anthropologist Dr. Maurice Bloch calls the "transcendental social."[3] His research suggests that one of the defining characteristics of our human species is our ability to *imagine.*

It's theorized that we humans developed this ability during the Upper Paleolithic period. It was a big deal. We began living a large part of our existence interacting with the world based on the imagined meaning, rituals, and roles we ascribed to our surroundings and to one another. We began to live a lot of our days in this space. We began to dream up social structures woven with patterns of utility and meaning. Rather than responding to one another based solely on what our senses told us in the moment—an experience requiring constant adjustment—we also began to interact according to essentialized roles. For instance, an "elder" could mean something specific and be a "role" beyond our particular temporal experience of them as an individual. This allowed us to live within imagined worlds in addition to our direct sensuous experience. In other words, we became creative.

If we can imagine our current paradigm, we can imagine others.

Of course, our reality doesn't *just* take place in our imagination; we embody it, and others do too. And, in an embodied reality in which humans

---

3. Bloch, "Religion."

have created systemized inequity and violence, the risks of re-imagining are not distributed equally—whether it's the end of a long, two-shift workday and the kids need to be fed, or one risks actual death at the hands of the State. We've dreamed up and realized systems that make *un*dreaming them an often unaffordable luxury, or the equivalent to wearing a dangerous target. It's profitable to keep us collectively unaware of the fact that we can imagine beyond today's instances of "is-ness." We are at once immensely powerful, immensely trapped, and everything in between.

&

These times have stripped the emperor of any remaining dressing. In the middle of our cavernous crisis—without direction or mandate—community businesses, organizations, and groups have stepped up to care for one another. They are trying out different ways of working together: case-by-case rent cancellations based on need; transparent and collective pay restructuring for resiliency; or continuing subscription payments on supplier contracts even when goods can't arrive, among others. "The way things are" has been disrupted. Crises do that. But the deeper truth is that we have *been* in crisis for a long time. Our condition is chronic.

Every day our organizations and businesses face profound questions about how we should treat one another. How we answer those questions is at the core of what perpetuates the ethics of our existing economy, or, conversely, works to reimagine our systems. Every micro-decision either creates the worlds we aspire to build, or contributes to the continual recreation of the world we have.

But what if we all woke up tomorrow and had complete "economic amnesia?" How might we choose to enact our economy? Could we end up in places completely different than what we know today?

Perhaps we would view housing as a utility instead of a commodity. Or, having forgotten the invention of "private ownership," maybe we would decide to manage resources differently, seeing them as common to more than just some of us. Perhaps the "one percent," no longer remembering "their" wealth, might abandon their fortresses and join the rest of us. After all, what would there be to protect, to reinforce, or to fear?

When we exit the pandemic, will we remember that there is no "that's just the way it is" and see instead there is only, "that's just the way we're imagining it right now?" Will we remember "that's the way we've always done things" is no justification for "therefore, we must continue on" with ways of being and doing that harm and cause inequity? Within *the world-that-was, shattered*, how can we support our community businesses, organizations, and groups to midwife very different dreams?

Crisis reminds us that imagination is perhaps more accessible to us than language. More like breathing than walking. We certainly don't have to get an MBA or go to college to learn how to imagine. It's already a part of who we are.

If going "back to normal" is actually a chronic nightmare, then what might the prayer be?

~

I believe one of the most fertile dream spaces is that of microeconomics: the decisions individuals, groups, organizations, and businesses make. As much as we are trapped and inherently complicit, these everyday spaces provide possibilities for surprising latitude. The changes may seem small at first: choosing to have a conversation with a co-worker differently; rewriting a contract with values of mutualism and trust; suggesting a co-creative approach during a meeting; structuring a salary amount based on need, or even potential, instead of the lowest bar the *market* dictates; or dismantling colonial notions of who holds the crucial knowledge and expertise.

From what might seem inconsequential, these intimate moments start possibilities spinning in ways that actually weave different futures. By training ourselves to believe in and remember the diversity of possible economies, their humble beginnings appear before us. If we know doorways exist, our brain looks for them.

I agree with J. K. Gibson-Graham that the economy is a space of ethical action. An economy is made up of a shared set of decisions imagined by all of us, human and non-human alike, in a relational fabric. It is partially an expression of our morality: what we choose to prioritize as good or notice as worthy, and what we measure or don't measure. Microeconomics is therefore also the domain of micro*ethics*: the common, deceptively routine juncture points within the internal workings of groups, teams, and enterprises that are indirect, subtle, and sometimes unconscious moments of ethical *choice*.

Although we are often told economics has an underlying logic, even a set of *natural principles*, Gibson-Graham rightly note that "our economy is the outcome of the decisions we make and the actions we take."[4] Contrary to how we often use the term, values are not aspirational; they are habitually expressed.[5] Our actual values, or those of a business, are the ones we

4. Gibson-Graham et al., *Take Back the Economy*, xiii.
5. Robb, "Values."

reinforce day in and day out. Values specify a general direction of continual travel, and they are the building blocks upon which we construct the imagined world of our economies.

In my work, I often see an imagination deficit around ethics when it comes to microeconomic choices. For example, organizations replicate the same injustices internally that their missions serve to counteract, or businesses receive praise for anti-racist rhetoric while continuing to replicate white supremacy within their structures and choices. The way we work is never neutral. It is always imbued with the ethics from which we select.

I am constantly asking myself, "What are the ethics from which I am currently selecting? What are the values I am expressing?" Not: "What do I *believe* in?" or "What do I feel *should be* in the world?" That part is easy. I'm asking instead, "What am I *expressing* in the ways I operate in this world?"

An economy is created, recreated, and again infinitely recreated, therefore presenting a cornucopia of chances to choose options as infinite (or as limited) as your imagination. There is the implicit social contract we each signed when we took our first breath and entered into interdependence: *My decisions matter. What I do—or don't do—creates ribboning pathways, encoded with the values I express in doing or not doing in a given instance. I affect emergence.*

Our businesses, teams, workspaces, and organizations are filled with so many more ethical juncture points than we've been conditioned to see. Often unaware, we make decisions by *not making them* and continue building castles from the sands of "the way things are," rather than from the way we want them to be.

This is not an argument for placing the onus of change in the hands of the many, while the few continue to extract from us. Our systems are imagined with intent. We of course didn't get into this deadly briar by accident. Even though imagination is innate to every one of us, the system in which we currently operate inequitably distributes the ability to turn imagined worlds into real ones.

Yet, every one of us is making the path as we walk it.[6] *How* we do things imbues the outcomes with their textures, tastes, and hues. The means are the ends. The future is just the pregnant present. Which means the *way* we work deeply matters. It may not be as sexy as full-scale policy change, but it is no less collective than collective action. We are in a continual dance of collective *becoming*. The degree to which a specific future is possible is directly equivalent to the degree to which we behave in accordance with that future in the present. Right now.

6. Horton and Freire, *We Make the Road.*

Bringing intentional, ethical choice to the realm of microeconomics is tremendously powerful. I believe it is a place where we can regain our footing and voice in the face of the intimidating scope and scale of what needs to change. I believe the *way* we work is one of the most potent and overlooked places from which shifts happen.

Living is a far more choiceful affair than the dominant terrain tricks us into believing. Underneath that pavement there are other *worlds,* teeming and overflowing with possibilities. Other presents. Other futures.

~

As of June 2020, seventy-five percent of people in the US didn't believe capitalism is working, according to a Just Capital report.[7] Yet, we help remake it daily in how we choose to build and participate in our businesses, organizations, and other groups. Just as it's difficult for us to imagine economies more diverse than the famed "ism's," it's perhaps even more difficult to imagine a business or organization separate from capitalism.

The pandemic has shown us firsthand that *we* are the economy. Amanda Janoo and Gemma Bone Dodds wrote in a spring 2020 blog that as we "stand still, the economy equally becomes more still. Our tendency to move, gather, and work together is fundamental."[8] In case there was doubt which lever affected the economy most significantly, we have found out. It's not corporations and wealthy so-called "job creators." It's all of us and our concert of micro-decisions. Microeconomics includes a whole set of underutilized levers—ones we already adjust every day.

In the months and years to come, the existing and the next generation of businesses and organizations will begin reconstruction. If we rebuild our businesses, organizations, and groups with the ethics of the dominant paradigm, we will create structures that again oppress, extract, and kill. Could we select from different ethics and express different values? I am interested in a reconstruction that doesn't just make marginally better what was already an intolerable state of affairs, but instead emboldens us beyond what we previously deemed plausible. What would it look and feel like to work in ways that are rooted in liberation?

We are told the economy only changes through big shifts, like those that require policy actions. This is the domain of macroeconomics. But what if that's only part of the picture? What if we can be ensnared in the barbed briar, and still make choices each hour and each moment from a set

7. Just Capital, "What Americans Want."
8. Janoo and Dodds, "Great Pause."

of values that are different from the set of values from which the current economy is derived? By picking one thorn out at a time, we also plant one foot in a different imagined direction.

Could we start to practice the everyday, rote realities of the future we want, *now?* Could we challenge ourselves to prototype the extent to which we can live out our microeconomics as though we've already achieved the macroeconomic shifts we so desperately need?

I believe the task at hand is not to *supplant* our existing paradigms for new, one-size-fits-all solutions. In our *undreaming* we must, instead, *unravel* the hegemony of "this is the way things are," leaving space once-again for multiple worlds to speak multiple truths. It is in those spaces that diverse, liberatory futures can emerge.

This is why I've been thinking a lot about probability. Not only the probability of COVID-19 existing on the skin of a peach at the farmers' market, but also the probability of "will we do it?" *Will we do the hard things, the reckoning, the repair, the midwifing?* In the aching of the now, I sense the presence of an insistent question: might reimagination become more probable?

They will say, *as they always say*, "There isn't time to get it 'right'!" Or "This is just how 'business' works," and "There's the bottom-line to think about." Or "What is the 'business case'?" I understand these might still seem valid, but only if one has been inculcated into the specific ideology of the dominant "now."

I find these rationalizations to be the rustling of paper ghosts. They can never stand up when held against the tests of who we *really* are, what we *really* want, or against the real constraints of our ecology. They don't belong in those spaces, spaces where our beingness—and that of others—is so poignant it's impossible not to be felt—gently squeezing someone's veiny hand for the last time as they lie in a hospital bed, or standing on the coast in the damp sand watching the ocean currents begin to communicate with hurricane winds. In those moments, we become like children reckoning with the abstraction of what we adults call *reality*.

∾

When I get tired, I start again in dreams—the fertile, prayerful parts of our realities. I see you there with me. On the back of the water buffalo, we weave the imaginal threads out of our bodies and into the purple sky. All the stars begin to twinkle with undulating potential. She talks to us. She reminds us of how impoverished of imagination we once were, of how it felt to starve that way, of the empty ache, of how little choice it felt like we had, and of how many somebodies tried to stop us from imagining into being the liberatory futures that now soak our days.

The water buffalo shivers as though just *remembering* brings back the rough winds and the high seas. I take the quilt down from the sky and wrap it around us. I whisper in her soft, oblong ear, "It's hard to even imagine."

References

Bloch, Maurice. "Why Religion Is Nothing Special but Is Central." *Philosophical Transactions of the Royal Society B: Biological Sciences* 363 (June 2008) 2055–61. doi:10.1098/rstb.2008.0007.

Gibson-Graham, J. K., Jenny Cameron, and Stephen Healy. *Take Back the Economy: An Ethical Guide for Transforming Our Communities.* Minneapolis: University of Minnesota Press, 2013.

Horton, Myles, and Paulo Freire. *We Make the Road by Walking.* Philadelphia: Temple University Press, 1990.

Janoo, Amanda, and Gemma Bone Dodds. "The Great Pause." *OpenDemocracy*, April 3, 2020. https://www.opendemocracy.net/en/oureconomy/great-pause.

Just Capital. "Survey What Americans Want from Corporate America During the Response Reopening and Reset Phases of the Coronavirus Crisis." Accessed August 8, 2020. https://justcapital.com/reports/survey-what-americans-want-from-corporate-america-during-the-response-reopening-and-reset-phases-of-the-coronavirus-crisis.

Partially inspired by The Henceforward. "Episode 26—Meditating on the Elsewhere." The Henceforward, November 12, 2018. Podcast, 16:41. http://www.thehenceforward.com/episodes/ 2018/11/12/episode-26-meditating-on-the-elsewhere.

Robb, Hank. "Values as Leading Principles in Acceptance and Commitment Theory." *International Journal of Behavior Consultation and Therapy* 3.1 (2007) 118–22.

14

The Income Guarantee

NATALIE FOSTER

COVID-19 OPERATES AS A social blacklight. We are seeing, with sharper contrast, what has been there all along—both our fragility and our strength. We are deeply interconnected. This truth rings out despite the myths and values of independence that have been peddled by the "pull yourself up by your bootstraps" ethos. The interdependence of our economies, our elections, and our bodies is coming more clearly into view.

This pandemic blacklight is highlighting the incredible economic pain experienced by many American families. As I write these words, over forty million Americans are out of a job. Achingly long lines of cars snake around parking lots with families hoping for a few bags of groceries from already over-stretched food pantries.[1] At the same time, we have marked out socially distanced squares inside other parking lots in a twisted attempt to keep our unhoused neighbors safe.[2]

I don't imagine that any of the folks in either of those parking lots need this blacklight to know what was going on. They already intimately know the realities of economic inequality. Even before the coronavirus pandemic, the average American family didn't have an extra $400 on hand to cover an unanticipated expense. So how do we expect people to make ends meet

1. Conlin et al., "US Food Banks."
2. Koran, "Las Vegas Parking."

when money stops coming in? Even if you did qualify for the insufficient, one-time $1,200 stimulus check, that money was spent before it ever arrived. Unemployment benefits only get you so far. How do you make the impossible choice between paying rent or restocking the fridge?

While we are all navigating the impacts of this pandemic and what it means for us, it is clear that this profound social, political, and economic disruption spells unequal consequences. Longstanding economic inequalities play a significant role in a person's ability to deal with this new reality. Money creates flexibility, but without cash on hand people are forced into impossible choices. Obviously, some jobs were able to translate into a work-from-home environment. But others must continue to work their job at the local fast food restaurant without proper protection just to keep a roof overhead.

The blacklight isn't only showing us new contours of economic inequality, however. The spread of COVID-19 continues to underscore the deep racial inequalities that plague this country. Across the US, coronavirus is devastating Black communities with higher rates of infection, complications from the virus, and loss of life. Essential frontline workers are disproportionately Black, which puts them at greater risk of infection.[3] Black people are also more likely to have an underlying health condition that makes them more susceptible to complications. Structural racism has created barriers to economic security, including access to quality healthcare, food, and housing. This inevitably makes Black communities one of the most vulnerable populations to a crisis such as this. But again, the precariousness of Black life in this country is not new. It is as old as our founding. Racism is the pandemic we have yet to truly face.

As COVID-19 moves through Black communities, state violence and police brutality continue to take lives. In a matter of weeks, the deaths of Ahmaud Arbery, Breonna Taylor, and George Floyd were brought into the national conversation and added to the long list of Black lives lost to grotesque violence and overt acts of white supremacy. In response, thousands took the streets in protest, calling out police brutality and demanding justice.

Those who were sworn to "protect and serve" were murdering people—we saw the boot of the state literally on their neck. Braving risk of infection, protestors exposed, once again, this country's racial contract.[4] This contract enforces the rules and norms guaranteeing protection and privileges to whites and denies their extension to nonwhites. The unrest we see around us isn't happening because of some breach of contract. People

3. Godoy and Wood, "Coronavirus Racial Disparities."
4. Serwer, "Coronavirus."

are rising up because they were never included. The profound upheaval of this moment is calling into question so much of the American experiment.

Work Is not Worth

As the existing inequalities and economic hardships multiply across the country, the magnitude of need makes it hard to ignore the twisted contradictions of our current economic approach. This pandemic and its impacts highlight the places where we have made egregious missteps and outlined ideas that no longer work. Near the top of this list is the idea that our work is what makes us worthy.

We have been taught to believe that work—our contribution to the economy—gives us our merit. Research from Anne Price and Jhumpa Bhattacharya at the Insight Center for Community Economic Development underscores this idea:

> Our country was founded on the puritanical notion of hard work and sacrifice as necessities in life, and as a result, Americans are obsessed with work. We define ourselves through what we do for a living and pride ourselves in constantly being busy and working long hours . . . We're so obsessed with work, we have come to define full personhood and deservedness with those who have full time, paid work.[5]

Beyond income, we've tangled up our employment status with access to healthcare and the basic necessities of food and shelter. Instead of guaranteeing rights, we have built a social contract administered through employers who have worked over the last several decades to shift the risk back to the employees. As more and more people are losing their jobs, they're also losing access to health insurance and benefits—a situation Americans who have non-traditional work have already endured for a long time.

There is a fundamental tension between the idea that work gives Americans their value and how the stock market assesses their worth. That the stock market could be rebounding when we have forty million unemployed Americans underscores the extent to which the machinations of our economy have given up on the American worker. These blatant contradictions, and the deep inequalities they communicate, are bone-chilling.

What if we were to build our policies upon the radical notion that every human has worth, that every person is deserving of housing, healthcare,

---

5. Bhattacharya and Price, "Power of Narrative."

and income? What if we oriented our economies around "being" instead of "doing?"

Stockton, California: A Solutions Lab

Guaranteed income is a robust response to so many of these contradictions. We see the benefits of divorcing work from worth playing out in Stockton, California. Mayor Michael Tubbs, the first Black mayor elected to office at the ripe age of twenty-six, is running a guaranteed income pilot. He makes the case that poverty comes from a lack of cash, not a lack of character.

We knew providing a stable income was important pre-pandemic, but post-pandemic it is a lifeline for Stockton families. Guaranteed income helps these families weather the crisis with more resilience than their neighbors. I can't help but imagine how this approach could have helped millions more if it had been part of our economic approach across the country. What would it mean to have this kind of an economic floor?

Named the Stockton Economic Empowerment Demonstration (SEED), the program gives one hundred twenty-five families in Stockton an income floor of $500 a month for eighteen months. The income is unconditional. This means there are no strings attached and no work requirements.

Stockton is a city in the process of reinventing itself, with no shortage of challenges. It was the largest city to go bankrupt after the last financial crisis. It's a city that is representative of America, with a majority of the population being people of color. One in four Stocktonians live in poverty; the median income is around $44,000.

The families participating in the guaranteed income pilot provide compelling data for how cash transfers allow families to remain resilient in the face of a pandemic. In these families we see how no-strings-attached cash provides a way forward in economic uncertainty. That resilience is the promise of the SEED project in particular, and the promise of guaranteed income more broadly. To see the impact on people's lives, let's look at one recipient's story.

For Tomas, the outbreak happened as he was getting his security clearance for a job at the airport. As with many companies, the pandemic caused a hiring freeze and halted the application process at the airport. Suddenly finding himself out of work and without any jobless benefits, the guaranteed income became Tomas' sole financial fallback.

Tomas can't live on $500 or even $1,000 a month, but the guaranteed income is not meant to be an income on its own. It supports resilience. It works to stabilize the erratic ups and downs and to help people through difficult times in times of widespread destabilization. This is something we should all have, and our current political and economic moment makes it a possibility. We can make this kind of economic care and resilience part of our reality in the wake of COVID-19.

We are often asked, "What do people spend the money on?" The answer: people spend the money on food, general supplies, and utilities. But wanting to know exactly where the money goes misses the point. A more interesting question is: "What does the money *do*?" The evidence shows that meaningful psychological and emotional gains are embedded in providing people with the resources to take care of their basic needs.

Researchers Dr. Amy Castro Baker and Dr. Stacia West, who are independently evaluating the Stockton program, understand this point. Beyond tracking the basics of where the money goes, they are evaluating questions of wellbeing, including stress levels, hope, and feelings of belonging.

The two researchers define hope in the way that they ask the question: "Do you want to wake up in the morning?" Hope is about goals, pathways, and agency. As Dr. Castro Baker puts it, "Keep in mind—the opposite of hope is despair. Understanding how economic security is linked to hope is key. Many would argue that change and justice are simply not possible without hope."

They also ask interviewees, "Do you feel seen? Do you feel seen as a human being by institutions with power over your life?" If you don't feel like you matter as a human, your capacity for hope is limited. Hope is one of the most significant predictors of whether or not you will engage with positive and healthy interventions when capitalism has already spit you out, or when it has communicated to you that you do not serve a purpose. According to Dr. Castro Baker, "Hope is about saying: To what degree can a justice-based intervention such as a guaranteed income serve as a financial vaccine in a prolonged stressful environment with an unknown end?"

Early trends indicate that, with just $500 cash a month, people can move the needle on both hope and belonging. Surely this is a powerful return on investment, placing it near the top of Maslow's hierarchy of needs.

"Power Through" by Astro; the Economic Security Project and SEED worked with Street Art Anarchy to identify muralists who could make visual reminders of the history being created in Stockton. Astro, a French street artist, brought a fresh perspective to the project. His mural, dominated by a layered green pyramid, speaks directly and indirectly to Maslow's hierarchy of needs and looks out over an empty lot in downtown Stockton. Photo by Andrew Laubie

For those us who need a refresher, the base level in Maslow's hierarchy of needs is about physiological needs: food, water, sleep, and the like. The next layer describes the needs for safety: health, employment, and a sense of physical security. The third layer is about love and belonging: the feeling or sense of community and connection. After that come the fourth level, esteem, and then the fifth level: self-actualization. There is plenty room for critique of Maslow's hierarchical model. For example, it's very possible to have a strong sense of community and still be food-insecure. As my friend and the brilliant writer, Mia Birdsong, chronicles in her book, *How We Show Up*. Still, the mural prophetically points to the ways a guarantee of income impacts people's lives.[6]

6. Birdsong, *How We Show Up*.

Better Ideas Aren't Necessarily New Ideas

It has been heartening to see guaranteed income percolate into Stockton society as a possible response to the pandemic and the treacherous economic insecurity experienced by so many Americans. Giving people cash works. It's not a new idea; Dr. Martin Luther King Jr. also spoke and wrote extensively on the power of a guaranteed income. He was pushed in his thinking by activists such as Johnnie Tillmon, who played a critical role in the National Welfare Rights Organization.

Nearly fifty years ago Johnnie Tillmon, a welfare rights advocate, wrote, "The truth is, a job doesn't necessarily mean an adequate income. There are some ten million jobs that now pay less than the minimum wage, and if you're a woman, you've got the best chance of getting one."[7] Tillmon's words are still true today. Over the last several decades, low-wage work has only continued to expand as we have transitioned away from industrial jobs to work in the service sector. COVID-19 has made this reality even worse. The National Women's Law Center reported that in the first few months of the crisis, Black and Latinx women suffer the highest job losses.

As a result of the pandemic, more and more Americans experience low wages, income insecurity, degrading interactions with benefits offices, and lack of protections. While these have been the consistent experience of many Black Americans, the blacklight is exposing others to this harsh reality as well. The extraordinary economic disruption is now moving beyond the segregated zip codes, where it has lingered untended for far too long.

While the impacts of the coronavirus mean that more people now experience economic inequality, people of color have been advocating for economic justice for decades. As we look for solutions and policy transformations, we discover a deep bench of women of color from whom we can listen and learn. Solana Rice, Executive Director of Liberation in a Generation, points out that, "When we co-create a big, transformative policy platform that centers on the economic liberation of people of color, we confront and dismantle the economic oppression facing Black, Latinx, Indigenous people, and marginalized people of color. Ending the profitability of racism will, in the long run, be better for everyone."

We now have an opportunity and a responsibility to advance new rules with old roots in a meaningful way right now. Even in a short amount of time, we've made tremendous strides. When I co-founded the Economic Security Project, guaranteed income struck people as a far-fetched idea and even a pipe dream. But that just isn't the case anymore.

7. Tillmon, "Welfare Is a Women's Issue."

As we are reimagining the social contract, guaranteed income continues to gain traction as a real and viable solution because it provides dignity and resilience in the face of so much uncertainty. Inside the Beltway, a group of elected officials (Senator Kamala Harris, Representative Rashida Tlaib, Representative Bonnie Watson Coleman, and Representative Gwen Moore) form a kind of DC "Cash Squad."[8] Even before COVID-19, each of these lawmakers pushed for real legislation to advance a guaranteed income that puts money into people's hands. They know the meaning and importance of this economic floor for millions of Americans struggling to make ends meet.

These women have big bold visions for how to remedy the ills of income inequality in this country, and they are pushing forward big bold legislation to get us closer to the goal. They are joined by dynamic community leaders like Aisha Nyandoro. She is running a guaranteed income pilot—Magnolia Mother's Trust—in her hometown of Jackson, Mississippi.

The success of Magnolia Mother's Trust (MMT) further speaks to the potential of a guaranteed income policy. MMT provided $1,000 a month, no strings attached, to low-income Black mothers for one year. Because of this program, the mothers were able to pay off over $10,000 of predatory debt collectively. They achieved educational goals, paid bills, and bought supplies and medicine for their children. As for the intangible benefits of the guaranteed income, mothers reported worrying less, having better family relationships, and feeling hopeful for the future. What started as a pilot program with twenty moms in 2018 has grown to a cohort of one hundred ten moms as of March 2020. In this new phase, MMT will also open and seed a children's savings account for all participants' children. Through experiments like SEED and MMT, we see how a guaranteed income helps to restore people's dignity and sense of agency over their lives. These programs create a foundation that can withstand the unexpected and, with coronavirus, the unprecedented.

Years of work on income inequality has taught me that the power of a guaranteed income lies in those two components: the guarantee and the income. People deserve an income floor both now and into our uncertain future. The need to rethink our social safety net has been long standing; it is made even more urgent by the current pandemic.

While the coronavirus pandemic is deeply destabilizing and causes no shortage of suffering, that same destabilization also provides a space for us to rethink *who* and *how* we are together. Let's use this space to guarantee more for each other; let's use this space to redesign our economy with

8. Foster, "Meet the Cash Squad."

resilience in mind; let's use this space to make our social contract more inclusive and equal.

Over the last fifty years, neoliberal capitalism has promised us so many things, but resilience was never part of the offer. As we stand today in the shambles of a broken economy, we get to make new rules. This time, we must include approaches that are rooted in economic resilience. We can pass policies that support families weathering the unpredictability and inevitable shock of this pandemic now, or a hurricane in the fall, or the longer-term horizon of job automation. Why should we wait any longer to create a better world?

As we look toward the future, people are asking different questions. Across the country we see the pain of communities that have been attacked, marginalized, maligned, and otherwise left out. We're seeing the desperation of people who are looking for a system that works for them. They deserve a system that values *life*—Black lives especially—and not just labor. Moving forward, we don't have to repeat the mistakes of the past; we can learn instead from what we are seeing in this moment. We can and should listen to the leaders of color who have been calling our attention to these political and economic solutions for decades.

The uncomfortable truth is that there will be future pandemics—whether they are the result of climate change, job automation, or something entirely unforeseen. We will find ourselves here again. But let's make sure that, when that day comes, the economic realities of American families are in a very different place. Right now, we can begin to offer the economic resilience required to weather these changes and the turmoil they bring.

As a nation, there is still much reckoning to be done. But we can't be an anti-racist, just society until we have safe and stable families and communities. And for this we need cash. Let's first provide a guaranteed income to stabilize every family. Then we can get to the hard work of healing this nation together.

References

Bhattacharya, Jhumpa and Anne Price. "The Power of Narrative in Economic Policy." *Medium*, November 8, 2019. https://medium.com/economicsecproj/the-power-of-narrative-in-economic-policy-27bd8a9ed888.

Birdsong, Mia. *How We Show Up: Building Community in These Fractured Times*. New York: Hachette, 2020.

Conlin, Michelle, Lisa Baertlein, and Christopher Walljasper. "US Food Banks Run Short on Staples as Hunger Soars." Reuters, April 24, 2020. https://www.reuters.com/

article/us-health-coronavirus-foodbanks-insight/us-food-banks-run-short-on-staples-as-hunger-soars-idUSKCN2261AY.

Foster, Natalie. "Meet the Cash Squad." *Medium*, March 02, 2020. https://medium.com/economicsecproj/meet-the-cash-squad-b7ee9b516ad3.

Godoy, Maria, and Daniel Wood. "What Do Coronavirus Racial Disparities Look Like State by State?" *NPR*, May 30, 2020. https://www.npr.org/sections/health-shots/2020/05/30/865413079/what-do-coronavirus-racial-disparities-look-like-state-by-state.

Koran, Mario. "Las Vegas Parking Lot Turned into 'Homeless Shelter' with Social Distancing Markers." *The Guardian*, March 30, 2020. https://www.theguardian.com/us-news/2020/mar/30/las-vegas-parking-lot-homeless-shelter.

Serwer, Adam. "The Coronavirus Was an Emergency until Trump Found Out Who Was Dying." *Atlantic*, May 8, 2020. https://www.theatlantic.com/ideas/archive/2020/05/americas-racial-contract-showing/611389/.

Tillmon, Johnnie. "Welfare Is a Women's Issue." Vancouver Rape Relief & Women's Shelter. Accessed August 14, 2020. https://www.rapereliefshelter.bc.ca/learn/resources/welfare-womens-issue-johnnie-tillmon.

15

Humanizing Work

Co-operatives after the Age of Capital

John Restakis

COVID-19

Today, as the world struggles with the profound uncertainty and risk of COVID-19, it seems as if all the fault lines and failures of the human community have been exposed to full view. The pandemic is forcing the global community to address foundational issues in our social relations, in our politics, in our economic systems, and in our connections to the natural world.

These problems are not new. What is new is their scale and the cumulative impact of their effects. What is new is the impact on our psyches and our collective sense of where we are headed. When it is announced that eight individuals own more wealth than fifty percent of the world's population, we are in the realm of dystopian science fiction.

The sense of vertigo that we feel is especially acute in the world of work. If anything conditions our sense of ourselves, our place in the world and our prospects for the future, it is our work. The rise in inequality, the devaluation and dehumanization of work, and the disrepute of the systems that perpetuate these conditions have crossed an historic inflection point.

The uprisings that are shaking regimes from one corner of the globe to another are a clear sign that momentous change is underway.

We have had enough.

The Co-operative Movement

At the dawn of the industrial age, resistance to capitalism was almost universal. Capitalism's triumph was not the inevitable result of some natural evolutionary process. It required violence, deception, the suppression of dissent, and the full weight of the state to function. Resistance took many forms. Among them was the struggle to preserve the integrity and humanity of work. The rise of the factory system, and the destruction of the rural commons that this entailed, recast traditional social values into new conceptions of work and community. Today is no different. The struggle to see work as an expression both of personal life goals and as a contribution to the common good is an old war now being fought on new fronts.

The co-op movement was a central part of this struggle. It advanced a radically democratic vision of society and the economy. Co-ops understood work as integral to human identity. Our work, our labor, isn't just a commodity to be sold in exchange for wages; it's a means for realizing personal and social well-being. As democratic enterprises, co-ops provide the means to reclaim control over our labor, to democratize our workplaces, and to humanize our work.

A co-operative is an enterprise that is collectively owned and democratically controlled by its members for their mutual benefit. While trade unions seek to change the balance of power between workers and owners in the capitalist workplace, co-operatives want to change the nature of the workplace itself. Their aim is to democratize the workplace and the relations of production.

All co-operatives direct economic practice toward social value. They operate for the collective benefit of their members and the wider society. These values define co-ops and their relations of production—not maximizing profits. At the present moment, this principle has a profound role to play in two crucial areas: the rising precarity of work, and the reconstruction of social care.

Precarity

In the world today, there are over three million co-operatives providing livelihoods to ten percent of the employed population globally. That is more than the top one hundred multinational corporations combined. Some, such as credit unions, agricultural co-ops, and consumer co-ops, are among the oldest forms of co-operatives. Others, like platform co-operatives, are new forms that are responding to the rise of the disruptive new technologies that are radically re-shaping the world of work.

Over the last decade, working people have seen a squeeze on wages more than at any time in the last 150 years. Labor market deregulation, the assault on unions, advances in AI, and the rise of digital technology pose unprecedented risks to everyone whose livelihood depends on a wage. Specialization within repetitive tasks by humans is the defining feature of labor in the capitalist model. This is what makes labor both productive and replaceable in an assembly system. But this model is mutating. Human skills cannot compete with computers. Without social controls, machines will ultimately replace *all* forms of repetitive human labor. Technology accelerates the precarity of existing jobs and will lead to the elimination of most employment.

What this process illustrates is not just the disposability of working people—the vast majority of humanity—but the loss of any understanding of work beyond its utility as a source of profit. Capitalism's drive to reduce the cost of production has pushed the de-socialization and de-humanization of work to new levels. Work is intrinsic to our identities; when it is reduced to a commodity in the production process, it *devalues* our humanity.

Uber's use of technology is a classic example. When Uber launched its mobile app in 2011, it was couched in the feel-good jargon of the "sharing economy." This was a cynical marketing ploy that appropriated the ideas and values of the social/solidarity economy to deceive the public and bolster private profit. In the social/solidarity economy, social values like mutuality and sharing are integral to the organizations that produce goods and services for social benefit. Co-operatives are a classic example, as are charitable organizations, community associations, and non-profits.

Uber is not. Nor are companies like Lyft, Airbnb, TaskRabbit, Handy, Clickworker, and Upwork—all of which use online platforms to broker exploitative transactions between actual producers and consumers. These transactions are not limited to simple things like food delivery. Upwork is a temp service directed to skilled freelancers like engineers, architects, lawyers, and management consultants. The company has global access to ten million contractors, who use a digital platform to bid against each other for work. The platform fuels a global race to the bottom for highly skilled workers who are often paid half the minimum wage in the US. All this is managed by a workforce of a mere 250 people.[1]

The human service sector is not immune. SuperCarers provides on-demand social care, and Teacherin is an online brokerage platform for supply teachers. Doctors and designers and a host of highly skilled human service professionals are also now caught in this expanding gig economy.

Digital corporations use online platform technology to extract value via a "black box" system that blocks any direct relationship between producers and consumers, as well as between the workers/producers themselves. Corporate policies are driven wholly by what will generate the highest profit. The data that show how these systems affect workers and the wider society are proprietary and hidden from public scrutiny. It is a "sharing" model where nothing is shared except the declining conditions of work. The impact of these platforms on people, communities, and the global economy is profound and as yet uncharted. Neal Gorenflo calls these "Death Star platforms":

> With incredibly low costs, global reach, scientifically developed user interfaces, and massive funding, Death Star platforms have a shot at duplicating this kind of success in every major city and service sector around the world. The lawlessness, network effects, and focused power of these platforms are now trained on the human services that were once the purview of public services. Since information-intensive industries have been monopolized by giants like Google, Facebook, Apple, Microsoft, and Amazon, this is the area where venture capital is focusing its support for new start-ups. And it's paying off.[2]

In the age of Google, despite the hip imagery of Silicon Valley, the command-and-control business model still reigns supreme. Workers have no power, social costs are unaccounted for, and the public interest is excluded. In the formative years of the industrial system, workers established a collective identity and mutual interest based on a shared work experience. The centralizing dynamic of the factory model meant they lived and worked together; they shared a common culture. This also meant they could organize. These shared social relations and their accompanying values are disappearing with the de-socialization of work.

The Uber model for the gig economy has perfected the best of all possible worlds for capitalist business: capital controls the production process, workers/society are powerless to monitor or counteract the effects of this

---

1. Conaty et al., *Working Together*, 19.
2. Gorenflo, "How Platform Coops."

control, and the erasure of connectivity minimizes the power of those affected.

But what if these digital platforms were controlled by the people who use them? What if the producers and consumers had power over the value they create? Enter platform co-ops. Just as the earliest co-operatives sought to humanize the economy through the creation of producer-and-consumer-owned enterprises, platform co-ops repurpose digital platforms to benefit those who actually do the work and create the value.

Shared Values and Shared Ownership

SMART is a platform co-op serving self-employed people across Europe. It was established in Belgium as a platform to assist people in the arts to access shared services such as accounting, bookkeeping, and legal services. But the co-op soon expanded to address a range of needs that individuals working on their own were unable to provide for themselves. These included announcing opportunities for collaboration on joint projects; sharing work and meeting spaces; forming education and training programs; accessing loans, insurance, and pension plans; and advocating for policies that addressed the needs of a new kind of workforce. Working with unions and human rights groups, SMART was soon active in organizing collective agreements on behalf of groups like bike couriers and deliverers. The co-op utilized its control over digital technology to reverse the isolating and disempowering effects of private, profit-extracting platforms. SMART now has some ninety thousand members (and growing) in seventeen countries. It is part of a larger movement to re-embed social values in what is now the dominant model of capitalist production: corporate mining and manipulation of data through digital platforms.

Platform co-ops like SMART are at the cutting edge of a new frontier. The commodification and manipulation of personal preferences and behaviors on platforms like Amazon, Facebook, and Google is premised on the erasure of privacy. This is a necessary feature of their business model. Part of the solution is to transfer control over personal data to the individual users of these platforms. A more effective solution is to democratize these platforms through collective ownership and control by those who use them.

Co-ownership and co-governance of the new production systems radically alter the content and character of the workplace and its meaning for workers and consumers alike. Wikipedia is a good example. Knowledge creation isn't controlled for profit but is organized as an *immaterial* form of universal commons accessible to everybody. This application of commons and co-operative logic can also be directed to *material* production for social benefit.

In the opening weeks of the pandemic, the collective design of cheap ventilators through peer-to-peer production systems using 3-D printers offered a dramatic example of the commons at work. In just one week, a group of doctors, technicians, and students collaborated to design a crowdsourced ventilator, the OxVent, which can be produced from widely available parts for under £1,000. If those participating in the production include the wider public or governments, these products could be treated as public goods. This is the exact opposite of the case of pharma giant Gilead. They are selling the COVID-19 drug Remdesivir for $3,120 per treatment—despite having received over $70 million in taxpayer money to fund the research. Nothing illustrates better the contradiction between public welfare and private profit in our current system.

The question is: how do social values get designed into the ownership and control structures of production systems that can transform the current paradigm?

Automation threatens to make precarity and mass unemployment permanent fixtures of capitalist society. As consumer buying power is eliminated through precarity, low wages, and unemployment, the very basis of the global economy is undermined. This model is unfeasible.

We can foresee two possible responses to this scenario:

- The state becomes the guarantor of social welfare and subsidizes a workforce that has lost its productive function. This strategy is a last gasp effort to preserve a capitalist system that is socially and economically unsustainable.

- The system is reconstructed by re-socializing the meaning and purpose of work. This is revolutionary and requires a complete re-thinking of value and the ultimate purpose of the economy.

If it is anywhere that work retains its social meaning, it is in the field of social care. Social care is a *relational good*—a good or a service that is embedded in an actual relationship among persons. It is the relationship *itself* that carries value. Relational goods acquire value through sincerity and genuineness; they cannot be bought or sold. Friendship and caring are relational goods. They are things whose sale would destroy both their character and their worth. When social care is an instrument for something other than the relationship itself, such as profit, it ceases to be care. The regression of social care into a system for the provision of *social commodities* (goods translated into monetary value) is the death of care.

The decline of social care has been at least forty years in the making. Influenced by free-market ideas, governments sought to reduce the social role of the state and to allow corporations to sell public services for profit. Corporate platforms have accelerated this trend. They further commodify and de-socialize what is arguably the last bastion of humane work. If online labor exchanges like SuperCarers can drive down costs by forcing caregivers to compete and isolating recipients from caregivers, all traces of genuine social care will disappear. The profit motive will ensure that. The convergence of neo-liberal ideology, which reduces everything to its market value, and the erasure of human relationships, which drives the gig economy, will finally hollow out what remains of humane work in the caring professions.

The collapse of productive labor is set to create a class of people—eventually constituting a majority in the industrialized world—whose worth can only be reclaimed in the domain of human, but not market-based, relations. This is the field of the social/solidarity economy. It requires the emergence of a new kind of market—a market of social reciprocity. The outlines of this paradigm are already in view.

Social co-operatives are the most promising attempt to re-humanize social care in our economies. They first emerged in Italy in the late seventies. Caregivers and families teamed up to create social care programs that were owned and operated by front-line workers and the people they served. In 1981, the Italian state stepped in to pass legislation explicitly recognizing the central role of social co-ops in integrating and serving marginalized communities.

Today, social co-ops are at the leading edge of social care reform in Italy, as well as across a growing number of states in Europe and Quebec. Their success is based on their ability to ensure that social relations are the basis of the value they create. Co-operative ownership means caregivers and those they serve are able to design the kind of care that responds to the actual needs of individuals and the wider community—not the bottom line of corporations. If we can proactively enact public policy to protect humane work relations for human services, we can support these initiatives in our own communities.

Fureai Kippu Co-op

Fureai Kippu is a co-operative system providing care to seniors in Japan. The term Fureai Kippu literally means "Ticket for a Caring Relationship." This refers to the ticket or credit that is earned when someone volunteers their time helping seniors. It is a time bank system where members can earn time credits or points for the hours they volunteer providing physical care, home help, and emotional assistance to other care-dependent members. These credits are then registered by their co-op and saved in their personal accounts. Time credit holders can withdraw and use their credits to buy care for themselves or relatives as required. The system is composed of a network of local co-ops that track and reimburse volunteer time on the basis of these earned credits. Credits can also be sent to other locales where the services can be redeemed to serve friends or loved ones there.

The model is an important complement to state care systems. Governments at both local and federal levels have supported the system. Yokohama City, near Tokyo, successfully recruited thousands of volunteers into the system by modifying the scheme to allow members to exchange time credits for services other than elder care. Young parents, for example, can use credits to pay for childcare or other services.

Fureai Kippu shows that reciprocity and mutualism can be valuated in social as opposed to monetary terms. The model shows how a reciprocity-based system of community-controlled co-ops can work with state systems to offer an alternative to the privatization of what should remain *social* relationships of caring.

Fureai Kippu creates a social market for the production and exchange of social value. Labor is valued without dehumanizing or commodifying it. Its social nature is recognized and rewarded. Above all, Fureai Kippu shows how an alternative value system can be the basis for a new kind of market— a new kind of *economy*—if the institutions are in place to give it form and effect. The credit that is earned by helping others is a form of social currency based on reciprocity. It works because people accept and stand behind its value.

A Way Forward

Free-market thinking would have us believe that only transactions determined by the profit motive—the price mechanism—have value. But our world is full of value: everyone understands the value of clean air and drinking water, the value of a beautiful piece of music, the value of loving and caring relationships, and so much more. These ways of seeing and thinking about value represent an alternative worldview to the transactional market economy that reduces all values to money. Markets—as we presently understand them—disfigure and diminish us both personally and socially by failing to give worth to the things we truly value.

We do not have to cede markets to capital. This doesn't mean that markets for commodities have no place. It means that not all value is for sale. The things we most value belong to another scale. In a world where labor has lost its worth, this essential fact can be the basis of something new.

What if an economic system were based on these premises: that it's the social worth of an action that generates its value. That human labor that serves a social good—such as caring for others, or teaching, or creating art, or tending the environment—is acknowledged and rewarded accordingly. Indeed, what if people could determine what those social benefits could be through the control they exert in the enterprises in which they work or the services that they use? What if we could perform not as disposable, exploitable, and replaceable parts—as mere human capital—but as co-owners and collective beneficiaries of the value we produce in common? And finally, what if the choices we make as consumers, or as investors, or as citizens, are similarly rewarded in proportion to the social value we create, and taxed according to the social costs we incur?

This human-centered dimension of value-creation holds the key to remaking—and re-humanizing—the economy in a post-capitalist future. It is the means for re-connecting work to personal fulfilment and for restoring its function as contributing to the common good. Instead of alienating individuals from each other and their society, reciprocity-based work helps to build authentic community. Just as work conditions one's attitudes and connections to society, the valuation of co-operation and reciprocity in work also fosters the most needed of all revolutions: the revolution in attitudes and values that restore our connection to each other and to the world around us. Unlike the "bullshit jobs" that create no value and shore up a bankrupt economic system, it is work that would actually have meaning.

The co-operatives outlined here illustrate forms of non-capitalist work that are operating now, at the very point when alternatives to the capitalist model are so urgently needed. The democratization of work and the economy offers the precondition for reclaiming the social meaning of work. It opens up the liberating possibilities that personal agency and control can bring to the work process. But it also means a radical re-engineering of public policy: regulating the gig economy, promoting worker ownership, ending the power of monopolies, reducing inequality, redistributing wealth, democratizating banking and finance, introducing a guaranteed social income, and, ultimately, creating a social reciprocity market that values and rewards the production of social value.

This isn't magic, or utopian thinking. It's simply imagining how one set of values might be replaced by another as the basis for how we reward and value what we produce as a society. It means re-embedding the social value that once determined what a market should, and should not, do. Ultimately, it involves a political struggle to reassert the collective good of society over the private privileges extracted from an economic system that no longer recognizes—or even understands—that an economy that dehumanizes its populace and destroys its human and material foundations is not an economy worth saving.

The models and mechanisms for transitioning to an alternative vision are already at hand. The democratization of the economy is one dimension of this epochal challenge. The democratization of the state and the role of government through radical political action is the other.

References

Conaty, Pat, Alex Bird, and Cilla Ross. *Working Together—Trade Union and Co-operative Innovations for Precarious Workers*. Manchester: Co-operatives UK, 2019.

Gorenflo, Neal. "How Platform Coops Can Beat Death Stars Like Uber to Create a Real Sharing Economy." *Shareable*, November 4, 2015. https://www.shareable.net/how-platform-coops-can-beat-death-stars-like-uber-to-create-a-real-sharing-economy/.

FOOD

Everybody Loves the Farmers Market
by Lindsay Jane Ternes

16

Planting the Seeds of the Future

VANDANA SHIVA

Adapted from an interview conducted on July 9, 2020

THE FARMERS AND PEASANTS of India, particularly the women peasants, have been my ecological inspiration ever since the days of Chipko in the 1970s—this lovely movement where women came out to hug the trees and say, "You'll have to kill us before you kill the trees." I grew up in the mountain forests of the Himalaya. I saw the forests disappear. When I found out about Chipko, I was doing a PhD at the University of Western Ontario in Canada, but I took a pledge to volunteer for this movement. What really inspired me was the deep knowledge that the peasant women of my region in the central Himalaya had of biodiversity.

I had worked as a trainee in India's fast breeder reactor and nuclear system, and then I did a PhD in the foundations of quantum theory. That kind of high cutting-edge science makes you think you know more. It has a way of creating a kind of arrogance in you; it sends the message: "You've done a PhD in physics—you're superior." The science establishment has largely been cultivated as a patriarchal hierarchy. So I like to say now that I did my PhD in biodiversity and ecology by learning from the peasant women of the mountains in the Himalaya. Even though I received my doctorate in physics, my entire ecological journey has really been shaped by peasants, and especially by the women peasants.

India in the COVID-19 Pandemic

The COVID-19 pandemic has affected India differently from the West because India has always been, and still is, a big attractor of colonizers, ever since the time of the British East India Company. In my activism over the last twenty-five to thirty years of globalization, one of my duties has been saving seeds, because companies want to own seed. They claim they invented the seed, and are pushing GMOs in order to push patents. I say to them: "No, you don't invent the seed. Seed evolves, seed makes itself. Seed is a commons."

India has a very large population: 1.38 billion. In these last twenty-five years, corporate globalization has displaced a lot of farmers because that's how the system is designed—to remove farmers from the land. As farmers were displaced over these last decades, they came to do very precarious work in cities, in very bad conditions. The Indian government announced our COVID-19 lockdown on the 24th of March at 8:00 p.m., to be implemented by midnight. Half the poor population was suddenly left without work or food, and nothing was done to help them. No arrangements were made, and no time was given for them to get home to their villages. They started to walk, some for thousands of miles.

The saddest part of our lockdown, which has been more brutal than most, is that we have so many people in a precarious situation. Death can come not from the virus, but from the breakdown of paths that are already fragile. This is particularly true with the displaced farmers. Everywhere, industrial farming destroys family farms and pushes people to work on other people's lands, growing commodity crops. One young girl, a migrant farm worker, had walked about a thousand miles to get home. One hour before she reached home, she collapsed and died. A twelve-year-old girl. In another instance, a group of peasants—tribal, Indigenous people—was walking back. Since busses, trains, and all forms of transportation were shut down, and they were so tired, they lay down on a train track where they thought they would be safe. A goods train came that way, and it ran them over.

Gandhi said that the world has enough for everyone's need, but not enough for everyone's greed. In this beautiful land of Gandhi, this beautiful land of Buddha, the greed economy—globalization and deregulated commerce—has been allowed to run loose. In the year 2020, clearances have been given for cutting down forests for coal mines—*coal mines*! Our tribal peoples have been fighting this since 1995. But the pandemic lockdown and the emergency it created has allowed the opening up of something that should be locked down forever.

The Greed Economy and the Drive to Control

The ecological costs of the greed economy during the COVID-19 pandemic are very high in India. The social costs are extremely high. But I keep working with our communities, reminding them of our deep ecological roots, and how we can't afford for the current tragedies to create hopelessness and fear. In our world now I see a conscious attempt to create a very large class of disposable people, throwaway people. It is time to return to Mother Earth and give her our dedicated love and care. In that, our livelihoods are going to be created—in a new economy of care, a new economy of regeneration.

We have to be self-reliant again. We have to become food-sovereign again. We have to become seed sovereign again. And we have to become economically sovereign again. Again and again, Indian agriculture has been destroyed. Sixty million people died of famine in India under British rule. We had been so prosperous; at one time we were twenty-five percent of the world's economy. Then the East India Company came, collected taxes (which is always their way), and transferred $45 trillion—just imagine, $45 trillion—by extracting a grain tax from the peasantry, who were then left to starve. After independence, we bounced back. Given how we have regenerated agriculture after destruction again and again, I think the world can learn from India.

Most people don't realize that modern organic farming originated in India. The British sent a scientist, Sir Albert Howard, to improve Indian agriculture. He arrived and found the soil was fertile. He found there were no pests attacking the crops, even though there were lots and lots of insects. He thought, *I'm going to make the pest and the peasant my professor.* Howard wrote a book on what he learned, which I found during my research on the Green Revolution in 1984 after the Punjab tragedy. I wanted to understand: How did we go so wrong?

Howard's book, *An Agricultural Testament*, inspired the creation of the Rodale Institute and the Soil Association in England. Howard had learned two things from the peasants of India. First, he learned that India did not grow monocultures. We used mixed cultivation, or what people today call biodiversity. Second, Howard noted that Indian peasants practiced what he called the Law of Return—the idea that we have received from the Earth, and so we must give some of our organic matter back to her, to replenish her fertility so she can keep sustaining life on Earth. I call it the Law of Giving, or the Law of Gratitude.

As I said before, working in the scientific establishment creates a mindset of control. It creates an engineering mindset, the sense that you control the world. Part of the colonial mindset has always been that nature is not good enough. Now, big tech wants to improve us. They say that we have to upgrade humanity into a machine, that agriculture must be upgraded into digital agriculture, and that we need to farm without farmers. Real food is "unimproved technology" and needs to be upgraded to lab food. For them, nature is the problem, and being part of nature is the problem. They declare Earth and nature to be enemies. Google, for instance, made money in tech, and is now moving into the life sciences. They think they're going to determine what life is. The mission of their life sciences division named Verily is to "defeat mother nature." Their vision of life is an end of ecological civilization.

This is something I've had to tell a lot of young activists, because people are still trying to fix issues like climate change technocratically. They don't look at the full energy system. They look at a technology based on fossil fuels and say, "Let's change this little fragment in one part of our system and everything will be fine," without looking at the footprint of the materials they're using and all that heavy energy consumption. In the end, if you continue to use energy slaves, it doesn't matter where your energy comes from. You will continue to destroy the planet, you will continue to destroy work, and you will continue to destroy the ability of the Earth to heal.

Nourishment, not Pseudo-Efficiency

This technocratic pseudo-efficiency mindset has been used to create industrial agriculture, and to make it look like it's somehow producing more. In fact, it's producing less, while using more resources. It's producing less nourishing food, using—as Amory Lovins would put it—more energy slaves. I ask them: How can you call a system that uses ten units of energy to produce one unit of food energy "efficient," when ecological systems use one unit of energy to produce ten units of food? Especially when the ten "efficient" units produce one unit of bad, toxic food—nutritionally empty food?

COVID-19 has made people in India wake up to the links between food and health. Everyone was waiting for the vaccine. But it turned out that the people who did not get COVID-19 were the ones with high immunity. And who were they? The ones who weren't eating junk food. They were eating Indigenous diets: lots of turmeric, a lot of ginger, a lot of ashwagandha. As a result, there's been a waking up to the sophistication of the Indian diet. This is in line with Ayurveda philosophy and its foundational idea about diet, which is just as Hippocrates recognized: *let food be thy medicine.* Ayurveda says, *"Annam Sarvaushadhi."* Food is the best medicine.

The link between food and health—between the value of people and being food sovereign—is about rewriting the efficiency equation from a fossil fuel-driven equation (where we hide the fossil fuel and resource use) to one that reflects a care economy. Human hands are what give care. If we caress our babies, we do it with our hands. Giant machines can destroy the soil. A spray of glyphosate can wipe out biodiversity and the monarch butterfly. But hands only give love. Therefore, the only way to get food that nourishes you is to put hands back into taking care of the land. This means we have to stop pushing technologies that displace people.

In July, the United Nations Environment Programme (UNEP) issued a critical report saying that leaders are only addressing the symptoms, not the root causes, of the coronavirus pandemic. I've written in many of my blogs about how new infectious diseases are being created by invading our forests in order to grow things like GMO soya, which goes to make biofuel and animal feed and is creating hunger and deforestation. New infectious diseases (SARS, MERS, Ebola, Zika, the coronavirus) are coming from the destruction of nature, the natural world, and forest ecosystems. This is the pseudo-efficiency of agribusiness at work. It creates the infectious diseases that come back to bite us.

We don't need to invade the Amazon. We don't need to drive butterflies and bees to extinction. We don't need to exploit and dispossess farmers. Already, eighty percent of the food we eat comes from small farms. We can make that one hundred percent of the food that we eat. Ecological agriculture, practiced with respect and awareness, can produce enough food to feed all of us. It's time for us to come back to root causes. And the root cause is pseudo-efficiency, which drives the invasion of ecosystems and violates ecological limits and planetary boundaries.

There is a Cree Indian prophecy that says, "Only when the last tree has been cut down, the last fish been caught, and the last stream poisoned, will we realize we cannot eat money." At this time of the coronavirus epidemic, and of the disasters of industrial agriculture, we must realize that *food* is the currency of life. Food moves between species to nourish life. Money doesn't move between an earthworm and a plant, but nourishment does. Nourishment moves from the mycorrhizal fungi to the trees and the plants. That cycle of nourishment is the food web, and therefore food is the currency of life. We have violated food as living and food as nourishing, and this is at the root of chronic diseases. Over the last ten years, scientists have been discovering the richest part of our ecology to be the gut microbiome, with sixty to one hundred trillion microorganisms. I call it *wilderness within us*. Not only do we need diversity to protect the Earth, but we humans are ourselves walking diversity.

Against Caste, Hierarchy, and the Greed Economy

India's COVID-19 lockdown has been so strictly enforced by police power that no protests or social gatherings of any kind have been tolerated. But before the lockdown, the big issues on Indian activists' minds were religious discrimination (against Muslims, for example, in the new proposed citizenship act), women's rights, and most of all, the caste system. The caste system is one of the biggest forms of violence in India. It takes a vocational distinction, originally made by choice, and freezes it at birth. Then it turns a diversity of occupations into a socially constructed hierarchy. Not only are all working people—Dalits—put on the bottom rung, but violence against them has increased.

The rigidity of the caste system can be traced back about two thousand years, when Buddhism and Jainism were growing. The Brahmins (those who studied religious texts) were threatened and responded violently. After Buddhism disappeared from the land of Buddha's birth, the Brahmins reinterpreted the sacred texts to define work in "making" as lowly. Caste became hierarchical, and those whose work supports society were relegated to a lower caste. The British entrenched the rigidity further when they created a census. I am a child of the fight against caste discrimination. My parents were big fighters against that system and for our right to choose our vocation; they thus deliberately chose a name for our family, Shiva, that was casteless. So my name, Vandana Shiva, was created to get rid of their caste.

According to a little book by Swami Vivekananda, the great Indian spiritual leader and an important teacher of mine, the British East India Company created a kind of rule by traders, which meant the rule of commerce. Vivekananda recognized that, until we undo this hierarchy and establish a rule of workers—those who create and produce—we will never have justice, dignity, or peace in society.

We have to stop allowing chrematistics, as Aristotle call it—the art of money-making—to be treated as if it were the art of living. To do this, we need to shift from a greed-driven economy to one driven by the original roots of the word *economy*: taking care of our home. But we can't practice the art of living until we awaken ourselves to the fact that we are not a privileged species. We are not superior to other species, as anthropocentrism makes us believe. The claim that some human beings are superior to others of a different color, or a different gender, or a different religion is yet another expression of the illusion of human superiority over nature, which breeds this sense of being above those whom you consider lesser than you. To me, this is what the protests in America are all about—the idea of white superiority. Here lies the root of the violence that is destroying the planet. The least we can do for our children is to give them a hopeful future. To correct the damage caused by this illusion of superiority, we therefore affirm the reality of oneness: that we are part of one Earth, interconnected through oneness.

Co-creation and the Biodiversity of the Mind

On our farm at Earth University in Navdanya in Doon Valley, where we save seeds and practice ecological agriculture, we have six times more pollinators than in the forest next door. Six times! Our water level has risen seventy feet. We weren't deliberately conserving water or creating a sanctuary for bee colonies. But nature was doing it. A lot of people interviewed during the COVID-19 lockdown have perceived that nature is coming back. I say to them, "Isn't it a message to us, reminding us that all you have to do is stop putting the pollution in the Ganges? You don't need to spend millions so someone can make money out of taking the next step."

We have to give up the monoculture of the mind, the militarized and mechanistic mind, and move to a biodiversity of the mind. We have to remember that nature is alive and creative, and realize that our own biodiversity is connected to the rich biodiversity of an interconnected living planet. This has been my deep lesson over these fifty years: All you have to do is your bit. You do not have to wrack your brains to engineer the whole thing. If you stop putting vehicular pollution into the skies, the skies will be blue. If you stop all the noise, the birds will come back. We are not Atlases carrying the Earth on our shoulders. The Earth holds us; we don't hold her. We tread on the Earth. And because she holds us, our work is just to stop trampling her. To tread lightly, with love. To serve her. If we offer this service, doing it in our tiny little proportions (because we are a tiny little species, one among so many), the rest will start bursting forth. Think of it as co-creation.

New possibilities emerge the minute you shift your mind from a dead Earth to a living Earth. An ecological civilization becomes possible. Co-creation becomes possible. And extractivism, which creates less from more, becomes unnecessary. When we go from being masters to giving love and care, all of these issues that look like impossibilities—solving climate change, reversing insect decline, growing our food without destroying the Amazon—now become possible because we are doing the right thing for the Earth, with love and humility.

When we bring back our heads, hearts, and hands, get rid of hierarchy, and learn how important it is at the present moment to cultivate the future, then we realize: nothing is more important than taking care of the Earth.

Hope and Resilience from India in the COVID-19 Pandemic

For me, every outrage, in a strange way, is a call to cultivate hope. It was at a meeting in 1987 where I first heard agricultural chemical companies talking about owning seed. They envisioned a future in which all seed, all food, and all health would be under the control of five corporations. At that moment I realized that a dystopia of bio-imperialism was being born, and that I had to start saving seeds. That's why I started the movement Navdanya, which has created more than 150 community seed banks.

I work across India, focusing on about eight or nine states, saving seeds and doing chemical-free ecological agriculture. During the COVID-19 lockdown, I began getting calls from my sisters in Bengal with tiny little pieces of land, saying, *thank you for inspiring us to start our Gardens of Hope.* I began working with women on Gardens of Hope because I realized that no matter how small the land of a woman, she could cultivate fruit. I started them really to stop the suicides. Now, they told me, because everything was shut down, if we didn't have our gardens, we'd have no food at all.

India is full of small retail, with street vendors everywhere, but the vendors are being pushed out by the COVID-19 lockdown and Amazon taking over. Peasants were losing their markets and distribution systems. It is an additional level of social closure in the pandemic that I call *digital dictatorship*: you can't choose to spent ten dollars at the grocery store; instead, you are forced to purchase through the digital system so that rents and royalties can be collected (like in British colonial times). So, on the 25th of June, I connected the two-million-strong network of street vendors with women farmers. I said: let us now create solidarity, build circular economies, and make these relationships dense and strong.

This is how I cultivate hope, in this moment of enclosure of the human spirit, the human body, the human mind: by connecting our deepest humanity to what is possible. Slavery was fought. It was considered normal, and then abolition happened. Monsanto would have imagined—and did imagine—that seed would be their property in the future. Instead, our agricultural movements created the possibility of seed freedom. These last five decades have taught me that the more dense an ecosystem is with relationships, the more resilient and productive it is. That's what we are learning from nature, and are now also trying to practice in the economy.

There is a huge, intense pressure coming to awaken in us the courage and resilience to fully live into our Earth-being, our humanity, our oneness. Each of us will find our different reserves to cultivate hope for a future that is for all beings and all future generations.

17

Food for Healing

MIKE JOY

COVID-19 HAS HIGHLIGHTED JUST how precarious our civilization has become. It is a timely reminder of how temporary and insecure past human civilizations have been as they have arisen and then collapsed. While previous civilizations were geographically constrained and overlapped in time, our current scale is unique. Never has a civilization been so immense, extensive, and interconnected. The pandemic highlights how globalization makes us one big precarious, interconnected, interdependent population, and it especially showcases the centrality of food and water security.

Right now, we face an unprecedented storm of imminent existential threats. Climate and biodiversity crises threaten the life-supporting capacity of the planet, along with more human-centered issues such as antibiotic resistance, pandemics, and the inequality that directly threatens human lives.

Almost all planetary life support systems have either hit tipping points or are teetering on the edge. Frustratingly, these crises are not broadly surprising, unexpected, or unpredictable. They take us by surprise only in their details or by dint of our willingness to stay deaf to the accumulating warnings from relevant subject experts. They are the known symptoms of a civilization that has overshot its biophysical boundaries. We have far exceeded our growth limits. This was predicted by Donella Meadows and her team at MIT in their report *The Limits to Growth* using the early computers of the 1970s.

Recently, tens of thousands of scientists teamed up to put out a series of "World Scientists' Warning to Humanity." These warnings include everything from collapse of food webs and microorganisms to freshwater biodiversity declines. Their common conclusion is: "If human behavior the world over doesn't change soon there will be catastrophic biodiversity loss and untold amounts of human misery."

I have found it to be really helpful to think about the limits of growth with the concept of planetary boundaries. Researchers at the Stockholm Institute calculated safe limits for a set of life-supporting planetary processes like biogeochemical flows, ocean acidification, atmospheric carbon and methane, genetic diversity, biosphere integrity, and land use systems. Their study revealed that we have exceeded most of these boundaries.

Surpassing these biophysical limits could never have been reached without reckless overindulgence. We have been on a spree of fossil fuel burning to achieve the growth of almost everything human-centered. In the blink of an eye on the time scale of human existence, we consumed the easily accessible part of our fossil energy inheritance. We were bequeathed this priceless legacy of accumulated solar energy that was charged up over millennia. Our inheritance was like a geological battery, giving us fuel that provided for an amazing set of technological advancements. The problem is that we exploited it as if it would never run out. And no, we cannot continue current lifestyles without this cheap energy. Something has to give.

We could have judiciously used these fuel resources to create a sustainable future, but instead we squandered it. I am sure if humans are around in a hundred years, they will look back on this time the way we look back on the aristocrats of pre-revolutionary France. They will see us as the architects of our own ruin through a heedless orgy of criminal and suicidal indulgence.

Our most extreme transgressions are those related to food production: land use systems, freshwater use, and biogeochemical flows (particularly nitrogen and phosphorus). The core driver for exceeding these boundaries is the dominance of animal agriculture. Our global food system exemplifies the dangers of opting for unconstrained growth using non-renewable resources. We live right now in a world built by an energy trove that is on the decline. We can no longer afford to exploit these resources if we wish to have a livable atmosphere.

The decline in available energy has profound implications for the way we live and produce food. It's not that we're running out of fossil fuels per se, it's the fact that we have used all the easily obtainable fossil fuels. Now it takes more and more energy to extract a given quantum of energy. This means more and more energy must be found and extracted and processed to have the same amount of net energy. Similarly, all the easily extracted materials have been mined, so our current prospects require more and more damage and energy to extract. Stated simply, the amount of energy available for humanity is declining and will before long drop precipitously. In any case, we have to stop these practices for the sake of the atmosphere—let alone all the other excessive and exploitative environmental impacts that fossil energy allows us to commit.

The demand for energy, material, and food keeps increasing as population continues to climb. While the rate of human population growth has halved since the 1960s, the actual number of humans added to the planet per year is much higher now than then. We currently add eighty million people every year. This means eighty million extra mouths to feed and supply with extra energy and extra raw materials.

Meanwhile, we have declining amounts and qualities of land for each person on the planet. This is simply because the planet is finite. As we add people, the available land area per person declines.

Eating Oil

The biggest technological change that drove (and enabled) our colossal human population growth was a scientific breakthrough from early last century. It is the Haber-Bosch process. Named after the men who discovered this process, it is the method, powered by fossil fuels, that creates nitrogen fertilizer. The Haber-Bosch process is the industrialization of a natural process: whereas nitrogen is captured naturally from the atmosphere by some plants (pasture clover, for example), this method achieves it synthetically using fossil fuels (mostly fossil gas). The scale of the exploitation of this process has become so extreme that humans have completely altered a global biogeochemical cycle. Humans now synthetically produce more nitrogen than all the natural earth systems combined.

We are now trapped in a catch-22. Our current food system could not support our current population without the contribution of fossil fuels for fertilizer, processing, and transportation, but the decline of energy availability and the harm done by its use means we must stop using it. As an indication of just how far fossil fertilizer alone has driven us past reality, without synthetic nitrogen fertilizer contemporary food production and distribution systems could not feed more than three billion of the 7.8 billion people currently alive.

While the introduction of this "fossil food" enabled exponential population growth, it was not evenly spread across the globe. Of the current population of about eight billion, nearly two billion people are moderately to severely food insecure, and at the other end of the scale two billion people are obese or overweight (this latter figure has tripled since the 1970s).

The synthetic nitrogen fertilizer that enabled this human population growth has had enormous negative implications for the living world because of the extent that humans have perturbed the nitrogen cycle. Humans now produce more nitrogen synthetically than all the natural systems combined. The process of creating synthetic fertilizer also requires large amounts of energy and adds significantly to the energy footprint of food.

The Haber-Bosch process is damaging because less than one-fifth of the synthetic nitrogen applied as fertilizer ends up in the food. The rest leaks out to the environment in different forms—and most of them cause harm. Some of the nitrogen ends up in the atmosphere as nitrous oxide, which is three hundred times more potent as a greenhouse gas than carbon dioxide and is also the most ozone-depleting gas. The increased levels of nitrogen in the atmosphere can be measured just as carbon can; the preindustrial level was in the range of 270 ppb (parts per billion) and that has now risen to above 328 ppb.

Much of the remaining nitrogen that is lost from agricultural systems ends up in freshwater—either directly through the leaching of fertilizer, or indirectly through animal urine that passes through soils to water sources. This excess of nitrogen drives aquatic plant growth. Under certain conditions (warmth and sunlight), plant and algal blooms begin to grow in rivers, lakes, estuaries, and oceans. These blooms cause extreme fluctuations in the dissolved oxygen available to the life in all these bodies of water. The low points in these fluctuations are lethal to aquatic life.

There are now more than four hundred known offshore "dead zones" where one finds little to no aquatic life left because of lack of oxygen. These dead zones cover more than 250,000 square kilometers; they exist mostly off the mouths of rivers with high nutrient inputs. On top of oxygen depletion, in many places these algal blooms in rivers, lakes, and oceans can also become directly toxic to aquatic and even terrestrial life.

The increase of nitrogen from synthetic production has impacts not only on ecosystems but directly on human health, producing significant economic costs. A recent study showed that the environmental and human health impacts of synthetic nitrogen use in the EU was many times greater than the long-term benefits. Here in New Zealand, colleagues and I have published articles showing that the negative impacts of intensive farming either match or outweigh any economic gains. It is important to note that the costs to society are not being paid. They are *accumulating as ecological debts* in the atmosphere and water, and are thus left for future generations to pay.

In addition to the well-known impacts on aquatic ecosystems, recent research has highlighted the link between nitrates in drinking water and multiple negative human health outcomes (particularly colorectal cancer). Evidence is accumulating that links nitrates in drinking water with thyroid disease and neural tube defects. Worldwide, colorectal cancer is the third most prevalent cancer and the second highest contributor to cancer deaths.

In New Zealand, this link between nitrogen and cancer is emerging as a critical issue because most of our drinking water comes from groundwater, rivers, and lakes where nitrate levels are high and still rising. New Zealand has some of the highest colorectal cancer rates in the world. Within the country, rates vary significantly, but the highest incidences are in South Canterbury and Southland—both areas that have high levels of nitrate in aquifers.

Livestock's Long Shadow

The rapid industrialization and growth of fossil food production coinciding with the rise in the human population has had dire consequences for wild animals on the planet. To give a sense of the scale of this issue, ninety-seven percent of the mammal biomass on the planet is now made up of humans and the animals we eat. The reciprocal and equally startling fact is that wild mammals make up just 3 percent of the biomass of mammal life on the planet.

This domination happened very rapidly. The current ratio is almost the reverse of what it was before industrialization, and it has ominous implications for biodiversity. The extent and intensity of animal agriculture globally is only possible through cheap non-renewable fossil energy. The process of creating fertilizer is the second largest contributor to human-made greenhouse gas emissions after direct use of fossil fuels. Animal agriculture is also a leading cause of deforestation, water and air pollution, and biodiversity loss. Half of the sedimentation of waterways globally is through accelerated erosion due to livestock.

Intensive livestock farming is responsible for one-third of all pesticide use, half of all antibiotic use, and one third of all anthropogenic nitrogen and phosphorus losses to freshwater. Livestock farming occupies the largest share of usable land globally. This amounts to roughly one-third of the land surface of the planet. One-third of that agricultural land is used for animal feed and forage, and livestock expansion is a major driver of land use conversions (especially forest destruction). Finally, the health and welfare of farmed animals is becoming a controversial issue in developed countries, meaning that the social license of animal agriculture is being lost.

Over the last century, as agricultural food production has become increasingly dependent on fossil fuels, the energy efficiency of food production has declined. Although yields have increased, there has been a concurrent decreasing energy efficiency. Since the 1970s the fossil fuel percentage of agriculture has gone from forty-three percent to sixty-two percent. The US offers an extreme example: on average, for every unit of energy in food consumed, twenty-one units of fossil fuel energy were expended on fertilizer, transportation, packaging, processing, and distribution. Furthermore, much of the gain in yield from fossil energy was used to feed animals for the increasing demand for meat. If this weren't enough, the energy losses in growing animals for meat means even less energy efficiency.

In most of the developed world we have fallen into a progress trap—a food production system based on the mechanization and industrialization of earth systems. This destructive transition was driven by a surplus of cheap fossil energy and buoyed by an economic system that rewarded short-term financial gain and ignored environmental impacts. The myopic drive for economic growth has destroyed many of the self-organizing ecosystems that had co-evolved over millennia. Our industrialized food process replaced these ecosystems with energy intensive, fossil-fueled farming and processing systems dependent on mechanical and chemical intervention. While many touted this as great technological progress, we can now see that it was an illusion. Though we can feed more people, we moved from a system that was stable, renewable, and resilient to one that is fragile, harmful, and self-destructive. Our future food production systems must be based on these natural systems—it is the only way forward.

Solutions

To have sustainable food production in the future we must immediately break our fossil fuel addiction. Clearly, we should have started this transition a long time ago, but we still have time. We can implement food production models that close nutrient cycles—ensuring that nutrients are cycled at the farm scale. We must protect and nurture soils, not mine them. Urban wastewater systems must capture and cycle nutrients for food production, and so much more. The simple answer is that we must mimic natural processes.

To be sustainable, food production must be in balance. It shouldn't need the intervention or addition of anything that is non-renewable. We know how to produce food this way. Agricultural production systems known variously as regenerative, closed loop, or biological can feed all the humans on the planet without harm to the living world.

One thing is clear, though: animal agriculture will have to be drastically reduced to achieve this goal. As we saw earlier, the human population has been artificially boosted with energy stored over millennia—fossil fuels. To make up this shortfall, animal agriculture must become a much smaller portion of future diets.

We will never reach the demands of a zero-carbon future or succeed at the transition to sustainable food until we accept the need for urgent change. We must reduce our consumption and growth. This means measuring the things that matter to people, not to financiers. The only way forward is through some form of true-cost accounting—one in which all the externalities, especially those in food production, are accounted for. When we aim for a sustainable future, gross domestic product (GDP) can never be the goal.

At the most basic level, it's the growth obsession that must be challenged and done away with. The growth paradigm, while imbedded in almost everything we have and do, was only possible through fossil energy, and so it must end. We must accept the fact that we have already far exceeded our planetary boundaries. De-growth is our only possibility, and we must share this knowledge widely. If we do not manage our way down from our current extreme consumption, then the consequences are inevitable. As tens of thousands of scientists have warned us, that future holds catastrophic biodiversity loss and untold amounts of human misery.

Conclusion

Our current predicament does not depress me—and this was reinforced by the COVID-19 response—because I know that the changes we must make to survive will be positive. I realize that this insight comes from a viewpoint of privilege, and that our transition away from fossil energy must be accomplished in ways that ensure equity, but what we give up will makes us happier and healthier. It will be messy and hard . . . and yet only messier and harder the longer we wait. The lifestyles we must adopt for a truly sustainable future will inevitably be built around small communities. Our food, fiber, and energy systems will be organized around localization, community, and an economy of sharing. Without our fossil energy machines we will have to put more effort into almost everything we do, but we will be happier for it.

Like passengers on the Titanic who believed the ship was unsinkable, most people on the planet believe that our civilization will never go under. Like the Titanic also, there are not enough lifeboats for us all. The difference is that we are still (just) afloat. We can save the ship! We have the knowledge and capability. The solutions will come from the bottom up, not from government. It will take awareness and education, since change must arise through grassroots actions and civil society supported by science. This can be the new legacy for future generations and our planet. Let's seize the precious time we have to correct our course and foster a real and lasting future together.

18

COVID-19, the Industrial Food System, and Inclusive Justice

Eileen Crist

"Suffering which has not yet come should be avoided."

—Sri Patanjali

THIS VOLUME EXPLORING THE implications of COVID-19 is being put together amidst the still swelling pandemic and its historic ramifications for human and nonhuman worlds. As I write these words, worldwide CO-VID-19 cases have topped sixteen million and the global death toll has surpassed 650,000 people. In the United States, an epicenter of the crisis, more than 150,000 people have died. After partial re-openings of businesses and civic spaces in May of this year, infection transmission in the US is resurging, dashing hopes of a slowdown from warming weather.

Data continue to pour in, but it is clear that people are not getting sick and dying equally. The elderly are most vulnerable, but that is only the most obvious facet of the COVID-19 profile. According to the CDC and other sources, a disproportionate burden of illness, hospitalization, and death is falling on poor people and people of color. The CDC highlights certain factors to account for the populations most susceptible: crowded living conditions, work circumstances, and obstacles to healthcare access. Buried among the CDC's listed factors of higher prevalence of infection among the underprivileged—and mentioned passingly—is the most significant: "serious underlying medical conditions." For example, the website informs that by comparison to whites "black Americans experience higher death rates and higher prevalence rates of chronic conditions." Dryly stated, as one might expect of a government agency.

More pointedly, racial and class discrepancies in morbidity and mortality rates are largely attributable to what the food movement calls *food injustice*. The disparity of the quality of food available to socioeconomically stratified Americans is cavernous. In truth, most Americans across race, class, gender, and other divisions are forced—to one degree or another—to contend with SAD: the Standard American Diet, largely composed of cheap animal products, sugary sodas and candies, low-quality vegetable oils, refined grains, and all manner of processed and prepared foods. "Conventional," as the label goes, refers to foods laden with pesticides, antibiotics, hormones, artificial additives, and more. Conventional foods are notoriously cheap, and thus especially marketed and attractive to poor Americans.

The environmental justice movement surged in the 1990s when sociologists and activists brought to public awareness the practice of polluting industries locating their operations in the vicinity of African American and poor neighborhoods. The food movement has taken umbrage with an equally pernicious pattern: Underprivileged inner city and rural areas are where America's food deserts cluster—human habitats where fresh and nutritious food is unavailable without a long, usually motorized trek. On the other hand, sodas, beef jerky, industrial-corn sweetened candy, Budweiser, and Doritos are a dime-a-dozen, so that food deserts are also *food traps*. Moreover, fast food franchises target poor neighborhoods to site their businesses. Unhealthy and over time disease-causing foods are consumed across American society and the Western-influenced world. However, the brunt of bad food consumption falls on the underprivileged, which in the US includes racial and ethnic minorities, immigrants, poor people, the institutionalized, and children.

$\sim$

In the midst of the current pandemic, the murder of George Floyd by a Minnesota police officer sparked a civil-rights uprising across the US and beyond. The Black Lives Matter movement has shaken America to its core. Its large popular base is reflected in the numbers of people of all races participating in demonstrations. What has also been politically pivotal is the widespread social and media solidarity with this civil-rights rebellion and its confrontation with a racist US administration. It has become more evident than it's been in decades that the racial divide—rooted in a history of slavery and still stitched into structural inequities in American society—must be dismantled and healed. Urgent changes are needed not only in the US legal system, prisons, and police departments, but more broadly in ensuring equal opportunities for employment, healthcare, good living conditions, and education.

The COVID-19 pandemic has revealed another disturbing systemic pattern that resonates with Black Lives Matter. Who is more likely to sicken from the virus? Who is more likely to die? The answer is the disempowered people of America, in large part because of the low-quality food forced on them economically, geographically, and culturally (for example, through targeted advertising). Today we know that the main driver of the chronic "diseases of affluence" is low-quality food that has become conventional in the US and in Western-influenced societies. While socially pervasive, these chronic diseases chiefly afflict the poor and otherwise disadvantaged. Those suffering from diseases of affluence—"serious underlying medical conditions," in the words of the CDC—are the populations most vulnerable to COVID-19 (or to any infectious disease for that matter).

Diabetes, hypertension, stroke, heart disease, obesity, asthma, certain cancers, autoimmune syndromes, and emergent bowel disorders—these constitute the *chronic* pandemics of our time. People are suffering and dying prematurely of diseases that are overwhelmingly preventable by means of a wholesome diet and healthy lifestyles. But the people who are afflicted by chronic diseases are not all suffering and dying simultaneously as with the current pandemic. The chronic diseases of affluence of our time are purveyors of what analyst Rob Nixon calls "slow violence." These diseases do not make headlines: they lack the sensational and spectacle-rife qualities of infectious epidemics; and they mostly afflict the downtrodden and demeaned, whose lives are tacitly deemed not to matter.

Morbidity and mortality from chronic diseases, disproportionately plaguing the disadvantaged, are pervasive but backgrounded, pandemic but muted, socially caused but treated as individually fated. They are the outcomes of structural factors, but those factors are obscured, or worse, concealed, by blaming genes or blaming the victims. The diseases of affluence cannot be redressed with vaccination; they demand revolutionary change in how we eat and how we live. These preventable diseases are globally escalating pandemics causing immense suffering and demanding colossal expenditures of funds and resources. They go mostly unremarked for the flagrant injustice they epitomize against the bodies of the poor, the under-privileged, and the uneducated, as well as against the bodies of children and institutionalized seniors or inmates who have no say on what's fed them. Indeed, toxin-loaded, disease-causing food is an assault *against nature* in the form of the human body across age, class, racial, gender, or any other lines. Bad food is also an assault against human potential, by cutting lives short; diminishing the overall quality of life; dimming physical vitality, mental acuity, and emotional clarity; and claiming resources that could be invested in sorely needed services such as universal family planning and education.

⁓

Enter COVID-19, the revelator: it is overwhelmingly sickening and slaying those people whose body functions and immune systems have been afflicted by an iniquitous food system. Yet the plot thickens, for the food system also abuses the natural world and domestic animals; it is anti-ecological and anti-life. Thus, hierarchical stratification on Planet Earth is *one*. In this hierarchy—which here I bring into focus with respect to the food system—non-humans are decimated and driven to extinction, ecosystems and biomes are razed, animals are unspeakably maltreated, disempowered humans perform the tedious and unhealthy labor, and the poor carry the biggest load of food-related illness and premature death. Ultimately, though, everyone suffers and will suffer by a food system that pays no heed—let alone honor—to the most precious gift we all share: planetary health, animal health, and human health to the extent that it lies within our power to cultivate it.

While there exists a diversity of food-producing systems around the world, my focus is on the spreading modes of industrial food production—the ones that make the cheap food driving chronic diseases, nature destruction, and animal hurt. Industrial food producers peddle products that run roughshod over the human body, as well as over our relations with the more-than-human world. These production systems are callous toward nonhumans (both targeted and bystanders) and are massively implicated in the deepening erosions of life. They are food systems propelling mass extinction, climate breakdown, and global toxification—intensifying catastrophes that harbor penalties that will make the challenges of COVID-19 look like a cakewalk.

Industrial agriculture operates through extinguishing ecosystems, and even entire biomes, such as grasslands that are all but globally gone. Vast monocultures are imposed on the land to mine the soil—the very soil created by the organisms who are extirpated and displaced to access it. Industrial agriculture pollutes land, freshwater, estuaries, and atmosphere with pesticides and artificial fertilizers, which are deployed as fixes for the problems that the agro-industrial model itself generates. The resulting pollution has produced hundreds of marine dead zones around the planet, while today even insect populations are collapsing, triggering the downfall of species that depend on them. Synthetic fertilizers contribute greenhouse emissions, and just manufacturing them yearly produces as many greenhouse gases as all US homes. Industrial agriculture claims eighty percent of the freshwater humanity appropriates, critically endangering freshwater biodiversity. Globally, freshwater populations of creatures have plummeted, with an unknown number of extinctions having occurred and more imminent. Where tropical forests are giving way to beef, soy, palm oil, or sugar operations, industrial agriculture is directly extinguishing lifeforms in exchange for mass-produced, internationally traded raw materials funneled into the industrial food system. Much of what comes from industrial agriculture is turned into animal feed for confined operations (more on that below), biofuel production, artificial sweeteners (implicated in rising rates of diabetes and obesity), and added ingredients in processed and packaged foods for the global middle class.

The impact of industrial fishing, in the service of mass-produced cheap fish, is one of the most ignored atrocities of our time. That impact is a consequence of a food-production system unaccountable to relations with the denizens and ecologies of the global ocean. Think scale: ninety-five percent of the ocean is open for fishing, much of it with mega-technological gear and on-board fish processing machinery. Cheap fish—available anywhere, anytime, and in any amount in the developed and increasingly the developing world—has exacted an appalling price. Eighty to ninety percent of the big fish are gone; former prodigious abundances of forage fish are no more; all species of sea turtles are decimated and endangered; seventy percent of seabirds have disappeared; many marine mammals are declining or barely hanging on; trawlers have demolished coastal seas, continental shelves, and seamount habitats. With fish and other marine life among the world's most subsidized and traded "commodities," the norm of cheap fish is spreading. What's more, chemical and plastic pollutants—many of them directly connected to industrial food, such as pesticides, fertilizers, fishing gear, and supermarket bags—are entering the marine food web, and that pollution is also seeping into human bodies.

Not just cheap fish, but cheap animal products in general are the norm. None are more egregiously extracted than those coming from Concentrated Animal Feeding Operations (CAFOs). CAFOs epitomize the severing of relationship. Domestic animal breeding programs have been accelerating, since the 1950s, to engineer bodies that grow faster, make more products, and churn out bigger litters. Lifespans have been minimized. Life processes, like egg laying and milk production, have been maximized. Whatever interferes with production—tails, beaks, scrotums, male chicks, or spent dairy cows—is liquidated. Mother-offspring relations are undone. Natural rhythms of sexual reproduction have given way to the depravity of clock-work artificial insemination. The feed that streams into CAFOs (after being laced with antibiotics and other additives) is grown on industrial landscapes saturated with herbicides and fertilizers, in the stead of grasslands and tropical forests, and including a sizeable portion of the global fish catch. The flesh of tortured animals, contaminated with unnatural substances, is turned into human food (though not for the well-to-do, who are parlayed organic, artisanal, and boutique fare). By their very makeup—overcrowded, genetically uniform animals force-fed copious antibiotics—CAFOs are incubators of perilous infectious diseases. Like COVID-19, such diseases are far more likely to assail the underprivileged, who are forced to eat the chronic-disease causing refuse streaming out of the places that are also brewing infectious pathogens.

~

Injustice pervades the entire world, human and nonhuman. From privilege to abjection, stratification constitutes a single pyramid. Its foundation is na-ture colonialism. Earth's biologically rich world, brimming with intelligence and splendor, possessing an extraordinary intrinsic and dynamic order, creative beyond comparison, and replete with sentient, aware beings—this world is transmogrified into "natural resources." Morphing life into resources enables the dominant socioeconomic order to spurn relation-ships with ecologies and nonhuman beings. What's made cheap and made without ethics in agro-industrial landscapes, in much commercial fishing, and in CAFOs is made by an exploited workforce composed of minorities, immigrants, minors, and other disempowered people (for example, prison inmates). The labor in those industrial sites is monotonous, underpaid, nonunionized, dangerous, unhealthy, and often traumatic or abusive. In brief, cheap food is made by colonizing the natural world, on the backs of the global proletariat, with cruelty toward animals, and at the expense of human wellness and longevity.

To create an ecological civilization it is of utmost significance for us to recognize and dismantle hierarchical stratification as a singularity. Deep transformation will respect the entangled wellbeing of people, animals, and planet. All human beings have the right to food that is wholesome, nutritious, and fresh; food made by preserving rich soils; food made with organic, ethical methods and in reverence for wild and cultivated biodiversity. The abomination of CAFOs must end. Food from the ocean can be consumed as an occasional treat (by those who still choose to eat it), acquired through the lighter touch of artisanal fishing. It is best to enjoy food grown locally or regionally, so that it is fresh and generates fewer or no climate-wrecking emissions. Importantly, the impact of food cultivated nearby is ecologically and socially far more transparent, for it is easier to invisibilize the destructiveness and inequities of foods imported from distant places.

In an ecological civilization, food cultures will vary from place to place, according to ecological affordances and cultural predilections. Yet a universal imperative in the service of planetary and human health is that the greatest proportion of nourishment come from a broad diversity of plants. By supporting the substantial lowering of domestic animal numbers, a mostly plant-based diet will free vast habitats for wild beings, biomes, and processes to resurge.

∽

When passion for inclusive justice—for planet, people, and animals—ignites the human imagination and becomes translated into action by communities, the slow violence of globalizing, chronic-disease pandemics will diminish precipitously. This will not only liberate people from unnecessary suffering and premature death. The parallel embrace of wholesome and mostly plant-based food will also spark a newfound recognition that health is a state of physical and mental wellness and stamina *far more profound* than just the absence of disease. Healthy food will also free up the skyrocketing amount of resources currently claimed by healthcare budgets. Once humanity has rid itself of preventable diseases, people will also be emancipated from the parasitism of big pharma, which—in classic disaster-capitalism style—makes a killing from the disease burden caused by the dominant food system.

With a robust immunity of equitably nourished people, when infectious diseases happen to move through the human population, they are far more likely to be sloughed off than cause illness and death. With the shuttering of CAFOs, one ticking bomb of infectious diseases will be disarmed. With the broad realization that the health of all—human and nonhuman—is intimately aligned, the violation of wild areas for industrial agriculture, poaching, mining and logging operations, and other profit-driven incursions will end. This will prevent infectious diseases, like COVID-19, that emerge through violence perpetrated against wild animals and their homes.

The conservation and restoration of forests, wetlands, grasslands, savannahs, rivers, lakes, and seas will create beautiful, connected expanses for the flourishing of all life, human life included.

One planet, one health. Humanity as one planetary citizen among millions, all equally deserving to be here and to thrive. We can choose to re-create ourselves as a modest, economically and demographically downscaled component of a biodiverse and rewilded Earth. An ecological civilization will accomplish far more than I've addressed in this essay. In the realm of food, an ecological civilization will respect justice and cultivate wellbeing for all. Does this sound idealistic? Perhaps. Yet it aligns with what author Ursula Le Guin evocatively called "the realism of a larger reality." That reality beckons, and now is the time.

EDUCATION

Survivor Justice
by Favianna Rodriguez

19

Education at the Edge of History

ZAK STEIN

I AM WRITING THIS essay in July of 2020 as an attempt to seed the future of education. Something will soon sprout from the inexorable institutional decay of modern mass schooling; the question is what. In the context of a global pandemic the *fragility* of large centralized school systems has been revealed. At the same time, there is overwhelming evidence for the *resiliency* of digital educational platforms. The tide was shifting toward using digital technologies as the basic foundation for new forms of education—now the tide has turned. The near-term results of this change could be disastrous, as the very foundations of enculturation and learning shift perilously close to chaos. It is possible that we stand at the threshold of the most profound transformation of education in history.

There is no future for schools as we have known them. Our world needs a new form of education—civilization depends upon it. While the days of schools are numbered, educational transformations and innovations are happening at Renaissance speed. After familiar forms of schooling disappear, education will remain. The future of education is going to be distributed throughout new and emerging digital information landscapes. As school systems continue to falter under the strain of unbearable complexities, we must be ready to abandon these older forms in the interest of education itself. This is the key to understanding education at the edge of history: new forms are coming into being.

I offer a brief overview of where we stand, and a way forward. Screens have beat out schools, and the pandemic finalized this transformation. The means of educational production have been transferred over to digital technologies. But these applications are not designed to be educational; they are designed to capture attention for profit. The only way forward involves a redesign of digital technologies, carried out in conjunction with a rethinking of schooling.

When Screens Replaced Schools

Society has been in crisis now—explicitly—for the better part of a year. Quarantines, school closings, and massive unemployment combined to trap the atomic family unit at home. Kitchens, living rooms, and bedrooms became classrooms, offices, and gyms. Information floods into and out of homes at alarming and unprecedented rates. With school and work miles away, and world historical events unfolding, the main educational effect is the informational ecology provided by the smart phone and computer. Most family members are "working" in one way or another at a screen, then switching chairs to "relax" in front of one.

We are awash in images, lost in the new media, between worlds in the midst of crisis. The total capture of attention by digital technologies and social media has now been achieved; the internet is everything during quarantines and lock downs. Yet we also know that it is an apparatus of surveillance capitalism. The pandemic moves us all deeper into the *enclosure* of awareness and communication within digital applications that are designed to reap a profit off our behavior. Views are like cash: this means all matter of enticements to click are put on the table. This internet, designed to addict adolescents, has won. Schools lost. It took a crisis to drive home what had long been the case.

With the advent of the smartphone, schooling was—more or less— permanently disrupted. Since its arrival, technology and advertising companies have captured the attention of the youth with ruthless efficiency. Digital technologies have supplanted schools as the main influence on childhood socialization and enculturation. For better and for worse, emerging digital landscapes of information outside of schools are educating the youth on a massive scale. Pop-cultural trends driven by capitalism have been competing with schooling since the 1950s; now, with the advent of digital technologies, schools cannot keep up.

The means of educational production have been turned over to new players. We are witnessing a transfer of power in the domain of who provides the main contexts, materials, and scaffolds through which education takes place. Today Google, Facebook, and a rotating host of other platforms own the means of educational production. These used to be in the hands of large international publishing houses, state-run schools, and universities; before that, education was in the hands of religious communities, tribes, and families. Today's changing of the guard places education firmly in the digital realm—which means it is securely in the hands of large private corporations. The implications of this turnover are unfolding around us. It's mostly occurring in the form of pollution and degradation within the informational commons. But it also takes place through the commodification of educational contexts in general, as the social system more and more reduces education to its monetary value.

(Re)Capturing the Means of Educational Production

Technological innovators make the tools that allow us to create and consume cultural content, and these tools are the means of educational production. Underneath the rising wave of digital technologies and media, a battle is being fought over who *controls and defines* the means of educational production. The technology that structures our "news feed" exerts more influence on us than the content of the articles and headlines. The aggregated effect of algorithmic curation is educational; the medium is the message. It's not only about what *we* are doing on social media applications; it's also (and I think primarily) about what these applications are doing *to* us. How are we being educated within them, over the longer terms of months and years?

"Data is the new oil" is now a mantra. The conquest, extraction, and profit centers of the capitalist world-system are no longer within mines and factories. This business of harvesting raw materials and turning them into commodities is now happening behind screens, sensors, and algorithms. The raw material is our behavior and awareness, which is extracted psychometrically based on our clicks, views, and purchases. This personal data is then rendered valuable to any number of interested parties—especially advertisers. "Advertisers" includes anyone who is interested in paying money to shape our thoughts, perceptions, and choices. They are not limited to people seeking to sell products; they also include a wide swath of political and ideological groups. The net effect is an unprecedented and invasive latticework of educational configurations that are more powerful in shaping the youth than any school system has ever been. As I mentioned earlier, the result has been a generalized eclipse of schooling; this is now accelerated by conditions surrounding the pandemic.

As children and adolescents (and adults!) scroll through endless targeted advertisements and curated feeds, they are being educated. This education is not merely the content of the "memes" and news stories they read. The "hidden curriculum" being taught within digital information landscapes presents itself as a drastic alteration of cultural forms and communication patterns. This hidden curriculum is the result of technological innovations ongoing for at least two decades. From the perspective of educational theory, the situation is something like the Manhattan Project—a vast covert effort to develop basic technologies that will change life forever.

In this case, the goal is to change the realms of culture, consciousness, and learning. Specifically, the goal is to capture the means of educational production and wield them in the interest of profit rather than the intergenerational transmission of civilization. This means reducing communication down to its most emotionally manipulative and psychologically addictive modalities: polarization, scapegoating, conspicuous presentations of self, provocative images, manipulative advertising, etc. All this is boiled down to the briefest duration and simplest structure. To get a sense of what that looks like, I have clarified at least six related features that combine to set the pedagogical atmosphere of most contemporary informational landscapes:

1. the inability to distinguish non-commercially motivated from commercially motivated information;

2. the related inability to distinguish honest information from intentional misinformation spread for strategic advantage;

3. decreased message length, increased message frequency, and the inability to track all message sources (i.e., information overload);

4. the absence of a shared overarching meta-narrative that could potentially reconcile conflicting information and perspectives;

5. escalating emotional intensity of information (due to factors 1–4); and

6. the normalization of weaponized language, such as lies, slander, censorship, and politicization (due to factor 4).

What happens when the primary modality of enculturation and education is characterized by these six features? It amounts to being raised in an informational warzone. Here everyone we meet is a strategic actor, and danger and confusion saturate the processes of identity formation. This is a complex technological, political, and economic problem. It's a problem that implicates the capacities and consciousness of people, especially children: their self-understandings, identities, fears, and hopes. Think about the extent to which screens mediate identity formation in our current culture. The result is an historically unprecedented opportunity for identity formation that is largely outside the bounds of normal reality testing, but within the bounds of high technology, hyper-communication, and hyper-stimuli.

We face an impending rift in the intergenerational fabric of the life world that is catastrophic for the continuity of civilization. To be cut off from functional, non-commodified, non-strategic communication is an untenable situation. We are living in the contexts of climate chaos, impending economic disaster, and an exponential increase in technological capacities, including the domains of weapons and biotechnologies. We need to be able to convene conversations and make decisions—in public, over time—and with some minimal, shared cultural coherence. Indeed, a culture that can come together in the midst of difference to learn about what is universally true for everyone is probably the definition of a coherent and healthy culture. This is another way of saying that education is the only answer to war that is not simply more war. Today the task of educators is found outside of schools: to reclaim the means of educational production and to redesign culture itself within digital informational landscapes.

The Future of Education Begins Now

I have written elsewhere about the fact that we stand poised between different educational futures—some good and some bad. Last year when I published my book, *Education in a Time Between Worlds,* the argument needed to be made that the large school systems built by modernity would soon be transformed drastically. Today we already see that process taking place.

In March, schools, colleges, and universities across the United States shut their doors and sent students home to learn in a new way. Each school dispersed into a distributed digital educational network, mostly unplanned and makeshift. School districts are figuring out how to make learning happen without school buildings, and colleges are finding a way to operate without a campus. As I have argued, much of the prior system of schooling is better left "switched off." We have the opportunity to end modern schooling and begin a new and truly digital era of education.

We are already seeing direct-to-consumer offerings, innovations in artificial intelligence-based tutoring systems, and technology-enabled pop-up classrooms beginning to reshape the educational landscape. In this time of opening and fragmentation we encounter sweeping science fiction-like vistas for educational futures. There are futures in which state schools have disintegrated into thousands of for-profit "EduShops" that sell software and remote tutoring and proctoring. There are other futures where massive online public schools teach millions of students exactly the same ideas in exactly the same ways—kids sit at home in front of state-distributed screens for hours on end. These are some of the futures we have to fight.

Although what I am saying may seem radical (that is, that schools as we have known them are dead), I am actually fighting for a future that embraces the accomplishments of public schools built by nation-states around the planet. These vast school systems of the modern world shouldn't be dismantled or shut down, nor should they be sold off to private enterprises (we are currently seeing the largest privatization of educational institutions in history).

Our great school systems need to be repurposed and redesigned—and now is the time. The school buildings themselves can be transformed into unprecedented institutions that are a combination of public libraries, museums, co-working centers, computer labs, and cooperative childcare centers. Funded to the hilt and staffed by citizen-teacher-scientists, these public and privately supported learning hubs can be the local centers of regionally decentralized pop-up classrooms, special interest groups, apprenticeship networks, and career counseling.

Giant schools built on the model of early twentieth-century factories can be gutted, remodeled, and reborn, metaphorically and literally, to create the meta-industrial one-room schoolhouses of the future—twenty-first-century temples of learning. Technologies can enable the formation of peer-to-peer networks of students and teachers of all ages, from all across the local region (or the world), without coercion or compromise. What enables these safe and efficient hubs of self-organizing, educational configurations are fundamentally new kinds of educational technologies. We can put almost unlimited knowledge in the palm of every person's hand.

The new sciences of learning are largely ignored or misused in the design of most educational technologies. The digital technologies we know are not optimized as *educational* technologies—not even close. For decades, research has told us that learning is optimized when it involves sustained interpersonal relationships, emotional connection, embodiment, and dynamically interactive, hands-on experiences. Based on what we know about the dynamics of learning, educational technologies should be bringing people together *away* from screens—*not* isolating individuals in front of screens. Technologies should help us customize learning and provide universal access to information through useful, well-organized, and curated content. They should not be the primary focus of attention or our main source of interaction and instruction.

Across the United States and throughout the world, many schools are being forced to patch together something like a digital system of education.

With makeshift stacks of existing educational technologies, we are experimenting on a massive scale with spoke-and-hub networks of decentralized mini-classrooms. During a "stay at home" order, every house in the country becomes a school, at least for a certain amount of time each day. That has not been the case since the era of the one-room schoolhouse (that is, not since pre-industrial education). Under the strain of social crisis, education retreats to its first and truest bastion: the relationship between children, their parents, and a network of concerned and responsible adults.

We should not think that keeping schools running now means having students sit at home in front of their screens all day. We must innovate, and we must do it radically and quickly. We can help communities and families realize that the power of education has been put back in their hands. Even though it may feel like a relief when schools reopen, there is a possibility that most learning will stay on screens. Our experience of a decentralized, resilient, and innovative digital education makes schools as we have known them appear obsolete.

But none of this is possible without reclaiming the educational commons and making good on the promise of the digital. I am talking about the difference between a pop-up classroom in a park and sitting in a chair watching a YouTube video. Compare a long conversation in real time to the asynchronous text-based exchanges found on platforms like Twitter. There is a stark contrast between actual embodied problem solving in the world and massive online multiplayer video games. To put it plainly, the choice is between reality and the screen, between the freedom of attention or its imprisonment.

Digital technologies could be designed to liberate attention instead of capturing it for profit. Bending history in the direction of a learning-centric (or human-development-centric) civilization requires that educational vision take precedence over business as usual. It is still possible to repurpose digital technologies for different ends, to recapture the best potentials of the planetary computational stack, and to avoid a catastrophic disruption of intergenerational transmission. It is not too late to save the very possibility of education from the clutches of total capture by capital.

As I have been suggesting here, the way forward involves the end of schooling and the end of digital technologies as we have known them. A future can emerge in which truly unprecedented educational configurations become the new normal. The vision of a decentralized network of education hubs outlined above offers one such vision. We can locate the seat of education within the community, outside school walls, and beyond campuses, engaging young people in the problems and processes of their community.

The secondary effects of the pandemic include fundamentally new economic realities and radical changes in the dynamics of labor markets. The "college to job pipeline" looks complex, to say the least—not to mention the school-to-college transition. Should the youth be sent back to schools as if they are being prepared to enter pre-pandemic higher education and labor markets? No. They should be released from this misconceived notion about the function of schooling.

Intergenerational transmission and education can be liberated from outmoded forms of schooling through digital technologies. But this can only be done if technologies are designed with educational value as the bottom line. Digital tools can enable people to safely find each other in actual embodied community—to collaborate, to learn, and to contribute to community problem solving. It may be that the only way out of the multifold crises cascading around us is to untap the wellspring of human potential. School-aged children and adolescents can actively solve many of the problems facing our communities if the right tools are put in their hands. Without this possibility, with only a vision of "returning to school," there will be a long, drawn out, and painful period of educational decline.

At the edge of history, we stand poised between a new dark age or a new enlightenment. The future depends on who controls the means of educational production.

20

What if We Looked at Our World as a Painting

Oren Slozberg

About ten years ago I was in an auditorium at a museum in Charlotte, North Carolina. Over one hundred teachers from different art disciplines were viewing a painting on a large screen. They had no idea what the title was, who painted it, or any other formal details about the piece. We started by looking at the evocative image of a woman sitting on a stool pictured below. We sat silently, taking this image in for about sixty seconds—she appeared to be next to a window with a figure outside. In the auditorium, every pair of eyes, every mind and body, was creating a story of what they were seeing on the screen.

July 7 by Frederick D. Jones, 1958.
Oil on canvas, 29½ x 24¼ inches.
Collection, The Minnesota Museum of American Art,
Elinor Brodie Fund purchase.

Over the next thirty minutes the group of teachers answered this simple question: "What's going on here?" Our conversation revealed the story each person had created by looking at this painting. The teachers were each asked to provide evidence for their stories with the question, "What do you see that makes you say that?" Their answers brought to light part of the process for how they make meaning. We heard what they noticed, how they interpreted it, and what assumptions they made.

What's Going on Here?

There could have been as many answers to this question as there were individuals in the room—or more. The woman was a mother who lost her son. She was a woman hearing a love song from her younger lover. She was Mary the mother of Jesus. She was a mother mourning her lost son. She was a washerwoman staring into space. She was a bride readying for her nuptials. She was a prostitute waiting for a customer. She was a lonely woman dreaming of better days. All these observations could be right—a *multitude* of right answers. And each interpretation was explicitly connected to some piece of evidence. Every line of thought from evidence to conclusion was made visible for all of us in the room. What was going on in the picture was any and all of those things. What was going on in the room was a process of dynamic, collective meaning-making.

This was a Visual Thinking Strategies (VTS) conversation: a simple facilitated conversation that engages viewers in making meaning from an art object. VTS is a methodology developed in the 1990s based on the work of psychologist Abigail Housen with museum educator Philip Yenawine. It has found traction over the years among elementary educators, art museums, and medical schools. After a VTS conversation in a new group, we often ask for reflections on the experience. Participants generally express amazement at having heard so many diverse perspectives on the same stimulus. So many stories flowing from the same source! Participants generally leave feeling like they've not only opened their minds to the art, but to each other. They have gained insight into one another's thinking and experience.

VTS conversations—and there have now been hundreds of thousands of them in schools, universities, and museums around the world—most frequently take place among students, where there is often permission for conflicting stories to co-exist. Adults also tolerate the various interpretations that flow from a VTS conversation because we understand art by its nature to be ambiguous and open-ended. Unless there is an art historian in the room, there is generally no major investment in getting a right answer. Resolution is not the goal. If the conversation leaves ambiguities, those unresolved ideas are actually welcomed as a sign of a generative and creative communal experience.

What happens if the stakes are higher? What if the artist were in the room and had a specific message they wanted to get out? What if there were a museum curator in the room who knew formal facts, such as who the artist was and the cultural context of the painting? They might put forth resistance to other interpretations. They might not like the suggestion that the grieving mother was a washerwoman.

VTS is just one example of open-ended modes of facilitated conversation that engage individuals in finding meaning together. As a method, it helps us to identify underlying assumptions and to listen to each other's lines of thought. It is a particularly potent one because it engages art and can bring out our creativity—it can make our hearts leap. But there are other systems of facilitated engagement that have similar effects, and still more are waiting to be invented and piloted.

VTS is the system I know the best. But the overarching question for us is: how can we find ways to talk to each other? Not to find consensus. Not to force agreement on complex issues. But to develop the skills of talking, listening, noticing our own investments and stories, and supporting our worldviews.

So, while discussing a relatively non-provocative piece of art gives people practice in noticing their own thinking without deep investment, how would their investments change if the discussion were not about a painting. What if we were talking about a public policy, or a healthcare decision, or an issue in their family? What if we were looking at an image, a video, or a text whose content challenged our livelihood? What if the overt meaning or imagery was in opposition to core tenets of a belief system that we thought underpinned our community?

Meaning-Making after COVID

Until March 2020, we each had customary ways of making meaning of the world around us. These ways of making meaning were tied to our life experiences, how we benefit or don't benefit from social and economic structures, the value systems of the communities of our birth and upbringing, and more. While there were already signs of system collapse—climate change and the breakdown of democratic processes, for example—individuals mostly were able to interpret their experiences in a way that conformed to the story they had already created or accepted. Individuals could do that without a conscious awareness of the worldview that shaped their meaning-making.

The spring of 2020 brought a pandemic and a popular anti-racist uprising persistently into the country's and the world's field of attention. While these factors have caused some thinkers and communities to look critically at the culpable systems, many Americans still found ways to make the new facts conform to their pre-existing stories. If data-heavy public health statements didn't agree with personal experience or with the words of a beloved and trusted president, then information was reinterpreted as either over-inflated or the result of a conspiracy. If one already believed—or wanted to believe—that our social order was fair and well-meaning, then the killing of George Floyd was interpreted as an isolated case of overzealous policing, not a symptom of a larger phenomenon.

We reach our conclusions about what we see based on ideas and experiences we already have, but without full (or sometimes any) awareness of what is undergirding that process. This is the tendency of the meaning-making process: to remain invisible.

Yet we are in a moment where change is not only possible; it is necessary. I'm not speaking about a specific policy or system change, although any given reader is likely to be an adherent of many. It is time for us to change *how* we think, individually and collectively. We are all very convinced of our rightness—and the absolute nature of that rightness.

But while outcomes may be evaluated by their kindness, harshness, or fairness, rightness is not measurable or observable. What we each call "right" is right through the lens of our experience. Something is right from the vantage point of our story. And our stories are sourced in many communal and individual factors. These include religious practices and traditions, capitalism and individual economics, race and other power structures, family history and circumstances, and the interests and effects of colonialism, physical ability, and availability of hope. Our individual experience is at the core of all of these axes of meaning-making—whether we benefited from or were wounded by any one of them. Our underlying stories impact who we see as important, what is important, what is still undone, and what (and who) is perceived as dangerous.

How do we recognize our investment in rightness—an investment that both energizes and polarizes us? We all know how difficult it is to change our worldviews. Even when facts disagree with our preconceived notions and we *want* our worldview to shift!

Changing thinking on a wide cultural level is a long and arduous process. We have seen this in many moments in our history where our public consensus has shifted. These shifts have involved protest, education, art, music, and political work, to be sure. But they have also involved a shift in the underlying personal experiences that inform the stories through which we view the world. This process is often slow and sometimes generational. Sometimes it's quicker than that. For instance, according to the Pew Research Center, in the fifteen years from 2004 to 2019, the US went from sixty percent of the population opposing same-sex marriage to sixty-one percent in favor. This change was not the product simply of rational deliberation. The change came from enhanced exposure—there are few Americans these days without a friend or family member who is gay and "out." The life of that loved one becomes part of the person's lived experience. It changes the story underlying their meaning-making as regards to the possibilities of love and marriage.

Abigail Housen, the developmental psychologist who articulated the theory of aesthetic development that underpins VTS, identified the range of meaning-making strategies we use when looking at art. She determined that most people (over eighty-five percent) use their own idiosyncratic values, experiences, and knowledge of the world to make meaning out of a visual image. What can be learned from that? What are the meaning-making strategies we employ when we look at a painting, a face mask, a police car, or a cultural or political development? How do we make visible to ourselves and others the sources of our interpretations? How do we become conscious of the personal and particular nature of our knowledge, values, and experiences (consider "The woman in the picture reminds me of my mother" or "I don't know any gay people")? And conversely, how universal are factors like the desire for love, safety, life, and belonging?

In her work, Housen was able to correlate aesthetic development—the ability to entertain different meanings for a piece of art—with the development of critical thinking skills. These are the cognitive skills that we use to analyze what we see and experience. They help us make sense of what we observe, reflect on where our thinking is sourced (through evidentiary reasoning), and build our capacity to communicate our thoughts to others. Housen's work documented how the act of engaging in creative viewing of art in a group setting can stimulate the growth of critical thinking. She also explored how critical thinking is developed and refined through VTS, internalized, and then applied to other meaning-making issues—including ones outside of the initial scope of inquiry.

So, what about matters of social and political policy? How might a process of open-ended inquiry (with or without a piece of art at the center) serve the goal not only of developing cognitive skills, but also of using those skills to engage in creative, open-hearted conversation across differences? What if we could feel the same clarity and self-reflection in a discussion of policy that arises in a group discussion of a painting? And what if we could foster the gratitude and compassion that grow out of hearing another articulate both their view *and* how they arrived at it—sharing the experience and values that brought them there.

But self-reflection and compassion are in short supply in our current political discourse. Our political polarization has made it so those who don't agree with us, whoever the "us" is, are seen instinctively as enemies. From Thanksgiving tables to Congress, we have given up on dialogue. We are polarized into constantly competing (and not always consistent) camps, as the pressing problems of our time continue to snowball. Our unfacilitated interactions become manifestations of our worldviews rather than explorations. Our dialogues are recitations of slogans and soundbites. Our political and social investments make it too difficult and risky to admit any uncertainty, gray area, or acknowledgment of the fears, hopes, and worries that underlie an opponent's understanding.

We cannot go on much longer this way—more and more of us experience this feeling. We need to evolve into a collective of individuals who—even when not agreeing—can listen, examine self-reflectively, and speak with humility from their subjectivity. Of course, people can and will organize to bring about their point of view. There will be political campaigns, marketing techniques, and social research. But there is something of tremendous importance, at the individual and small-group level, that needs to be addressed: our ability to think critically and to talk to each other.

How do we develop these skills? In this world of competing demands and frayed nerves, how do we learn to look at our own stories, see how they inform our current vantage point, and recognize how others' viewpoints are similarly rooted in experience, hope, and fear?

We are living in a complex world with wicked problems that are not easily solved. We all exist in systems that are linked and interact in ways that are invisible and multifaceted. Holding this level of complexity requires critical thinking and ample metacognition. There is a temptation to reduce complex issues to bite-sized problems in order to make them more accessible. However, in that reductive process we arrive at narrow and incomplete solutions to complex and interlocking problems. Consider, for instance, the illusion that if we stop using fossil fuels the climate will be okay. Or that, if we elect a new government, our polarization will end. This process of reduction fosters the belief that we can solve our existential problems with simple behavior changes or one-time fixes.

In the same process, we risk underestimating the magnitude and seriousness of the challenges we face. It is natural not to want to live with looming clouds on the horizon. But pandemics and social unrest are harbingers of more to come. Some impending risks we are aware of (climate change), but others are still remote or abstract (the unfolding collapse of the ecosystems of the Amazon). If we aren't aware of how we think and how we make meaning, we will not be able to discern the difference between meaningful solutions and fantasies. The challenges ahead are daunting. If we can treat the world as a piece of art (or many pieces of art!) with layers and hidden connections, then we can employ our creative juices and our critical thinking to face what is coming. This process encourages us to authentically and directly find our next steps together as a local and global community. So how do we bring about a culture and an era of critical thinking?

Let Us Imagine

I would like to envision a world where critical thinking is a key practice, like prayer, sports, and breaking bread. A world where critical thinking—not thinking alike—gives rise to a sense of richness and belonging.

Let us imagine what could happen if we cultivated practices of saying what we freshly "see," not what we habitually "believe." What would be possible if our eyes and ears were open as wide as our mouths? Or if we approached questions with less certainty and more curiosity.

Using a VTS methodology as an example (without investment that it be VTS), I could imagine a world in which:

- Libraries, churches, lodges, youth programs, YMCAs, county fairs, schools, and grassroots pop-ups offer open-ended conversation evenings—in person when COVID-safe, online when not. Begin with seemingly homogeneous groups (for example, people from the neighborhood, people of a similar age, or some other shared demographic).

- Begin with an image that feels culturally familiar to that group. Then move on to images that are more ambiguous or less culturally familiar, or images that might have a more charged resonance, like a painting of people in masks.

- Using a method of inquiry ("What's going on here?" "What do you see that makes you say that?"), along with creative facilitation methodologies (paraphrase, linking, distinguishing), allow people to engage in real-time practice in noticing what they tend to see and how they tend to make meaning. Notice how empowering it is to do this collectively.

- Let the conversation tease out the diversity of outlook and experience among participants that might not have been initially visible. Let the conversation teach how to hear others' experiences more open-heartedly.

- Create a network of those who offer such discussions. They can create opportunities online for alumni of the entry-level conversations to participate with a wider range of people. Don't push for any agreement—allow for greater understanding.

- Encourage the use of simple facilitation questions in relationships, families, workplaces, and organizational settings.

- Let an art discussion activity launch any difficult public conversation. How might city council meetings change if led off with the discussion of an image? Or with discussions of anti-racism? How might we be opened up, "tenderized," or habituated to listening and to noticing our own and others' story making?

- Promote local or national radio shows that spend an hour on three image discussions. People can connect online to see the image and discuss through call-in or chat features.

I have no illusion here about our becoming a nation or a globe of art-viewers and open-hearted *conversation-havers*. But we have abandoned our commitment to, and even awareness of, critical thinking. We speak without accountability.

We live in a time of great public and personal reckoning. We have an opportunity to think about what we can be doing better together. What do we need to do differently in the public sphere, and how can we do it differently in the personal sphere?

Accountability, true transparency, and openness—these are not apparent in our political discourse right now. Nor in nearly any civil discourse. We are in a moment of upheaval and struggle: environmental, racial, and economic. But in this national and global transformation, we must resist the polarizing paradigm of winners and losers predominant in our culture. We need to listen with compassion, cultivate self-awareness, and engage in communal meaning-making.

This is a tall order. But to become a more civil, just, and self-aware culture, we must practice. No one can do it for us—and it won't be so bad—but we can start by looking at a picture.

LOVE

We Will Heal Together
by Mira Sachs

21

Open Hearts, Open Minds

Jack Kornfield

"Our knowledge of science has clearly outstripped our capacity to control it. We have grasped the mystery of the atom and rejected the Sermon on the Mount. We have achieved brilliance without wisdom, power without conscience. Ours is a world of nuclear giants and ethical infants."

—Omar Bradley, Chairman Joint Chiefs of Staff

TODAY WE ENCOUNTER A new world. Yes, it's the same place, and yet somehow dramatically different. A world newly shaped by COVID-19. Now we wear masks in the grocery store. We say hello to loved ones from awkward distances and face a palpably uncertain future. National economies grind to a halt. Resources are marshaled. We do our part to limit the spread.

We are trying to meet the dire needs of this crisis, but our world is still mired in crises from before. Racism, environmental destruction, warfare, and global injustice continue to terrorize in the wake of this pandemic. The spread of this virus is a justification to relax environmental regulations and an excuse to act against civil liberties and democratic movements. As people take to the streets in protest, we are awakening to the racial inequities in our society.

COVID-19 has refracted the crises of our world through a new lens. We can see the suffering around us. We can listen to those calling out for justice, or even just for a breath.

But what can we do? *We will fail to meet these challenges today if we fail to engage our hearts.* The scale of the structural and systemic problems alone is stunning. We are now feeling them so acutely. The task before us is to hold these realities together.

Will we work for healing in the midst of suffering? Will we bind together to end cycles of violence? Today we face a world with familiar problems and new challenges. Let us embrace it with a compassionate heart.

Stop and ask yourself: Which offers a better future—a society that fosters greed, hatred, disrespect, and ignorance . . . or one based on generosity, love, respect, and wisdom? One leads to suffering, while the other leads to happiness and wellbeing. We recognize this. But when we open ourselves, our heart also feels the wounds of the world around us. This can be overwhelming, and the question "What can we do?" quickly bubbles to the surface. We feel the distance between this future of wellbeing and our current moment. How do we awaken the love within our hearts?

∾

We must pause and listen in a new way. This is not the way of more and more technology. The modern world moves with increasing speed and complexity, flooding us with tools for ever more multitasking. We marvel at the development of computers, internet, nanotechnology, space technology, biotechnology, VR, AR, and AI.

Yet these outer tools will not stop continuing racism, environmental destruction, injustice, and warfare. That is because all of these forms of suffering have their roots in the human heart. Technology can actually pattern our lives in false separation. In a world shaped by COVID-19, technology has been a crutch for businesses, a way to access entertainment at home, a resource for seeing each other virtually. Alongside the helpful tools that it offers, though, technology also has the effect of distancing us from the mystery of love and its beckoning whispers. It threatens to determine more and more of our lives as we grow in our dependence on its offerings. In our continual striving for connection, technology floods our lives with distraction and busyness. We check our email and our messages, all while looking up a recipe for dinner. These practices shape our attention.

This outer focus keeps us from cultivating the capacities of heart and mind that are necessary for healing our lives and our world. We must remember who we are. We are alive in this present moment. We are loving awareness itself. Our task in this time is to go deep, to plant roots, and to connect to our timeless existence and love. To be present. To remember who we really are: part of the great mystery.

As a Buddhist teacher and a psychologist, I believe we must learn to love in new ways. If we want to be a wellspring of transformation in our world we must match the outer development of humanity with inner development. We have innate capacities for wisdom, joy, hope, and love. The beautiful thing is, with attention and training these can be awakened. To answer the call of our time, we have to comprehend with our minds *and* with our hearts.

As a species, we have developed powerful tools and strategies for awakening the heart's wisdom that are a blessed part of our human heritage. Traditional cultures have developed many skillful ways to enhance compassion, lovingkindness, forgiveness, gratitude, joy, peace, and connection. Some of these practices, including many from Buddhist psychology, have been studied and replicated. In the past twenty-five years, neuroscientists have published some seven thousand papers and studies on the benefits of trainings in mindfulness and compassion alone. Researchers like Richard Davidson have amassed evidence that measurable changes to our brain and nervous system occur ten times faster when heart practices like compassion and lovingkindness are incorporated with mindful attention.

The heart knows this, and the body reflects this truth. We are connected in profound ways. Every breath we take has recently passed through the lungs of countless humans and a multitude of animals. It has dusted the tops of Mauna Kea and Mauna Loa in its journey across the vast Pacific and brushed the ruins of the Fukushima nuclear reactor. We exist in an interdependent web of life.

We come alive when our sense of separateness drops away. We remember walking in the high mountains, making love, losing ourselves in music, using sacred medicine, witnessing the birth of a new child, or sitting at a bedside at the moment of death when the gates to mystery open wide. This is the sacred reality of life.

⁓

The crises of our times require individuals and cultures to engage in a collective heart work: to grow our ability and willingness to see suffering and to speak out. To feel our grief, regrets, fears, longing, and confusion. But this is not the end of the story. For those who know suffering and its causes, they can begin to see that there is a path to the end of suffering. Generosity grows, clarity and love come alive, and the reality of interdependence leads the way. In spite of outer suffering our heart wisdom can express creative hopes, delights, joys, and all the goodness of life.

We have a choice. Epidemics, like earthquakes, tornadoes and floods, are part of the cycle of life on planet earth. How will we respond?

Will we be frightened by the suffering and double down on greed, hatred, worry, and ignorance? Or will we work toward healing with kindness, steadiness, and compassion?

This is a time for bodhisattvas. In Buddhist teaching, the bodhisattva is someone who vows to alleviate suffering and bring blessings into every circumstance. A bodhisattva chooses to live with dignity, courage, and radiates compassion for all. The time for this love is now.

Bodhisattvas are asked to receive the cries of tragedy and respond in love. We can hear them singing from their balcony to those shut inside. We can see them in our neighborhood caring for the elders. They are our healthcare workers and the unheralded ones who stock our grocery store shelves. They are in our own communities, in our circle of friends. I hold close my daughter and her first-responder husband who worked without protective gear for weeks. I hear from tired friends serving in food kitchens, with never-ending lines of parents and their hungry children in tow.

When we learn to quiet the mind and open the heart, we can sense our interconnection. In those moments we naturally want to attend to the shared welfare of all. An open heart brings benefit to every endeavor and domain. Our bodhisattva hearts overflow into care for the environment, business practices, education, even global peace. When we see with our heart, our family circle grows larger, until it encompasses the web of life itself. We realize the real question is not the future of humanity but the presence of eternity.

The work of the heart, though rewarding, is equally demanding. Whether we admit it or not, we are vulnerable beings. Heart work requires us to learn skills in healing trauma and managing fears. We need to spread widely, wherever we can, the skillful tools of compassion, nonviolent communication, and trauma work in order to promote individual healing and reduce the cycles of violence together.

To do this as a culture requires developing our capacity to hold the full measure of our humanity in the compassionate heart. Neuroscientists call this "expanding our window of tolerance." Without this wisdom, we blame society's ills on others, whether the immigrants, the Muslims, the communists, the blacks—always someone else. James Baldwin explains, "I imagine one of the reasons that people cling to their hate and prejudice so stubbornly is because they sense that once hate is gone, they will be forced to deal with their own pain." Our hate is a mechanism to keep us from facing our insecurities, our difficulties, our loneliness, to keep us from feeling the wounds of injustice around us.

What can we do? First, we can sit quietly, take a deep breath, and acknowledge our fear and apprehension, our uncertainty and helplessness, holding all these feelings with a compassionate heart. We can say to our feelings and uncertainty, "Thank you for trying to protect me. I am okay for now." We can put our fears in the lap of Buddha, Mother Mary, Quan Yin. We can place them in the hearts of the generations of brave physicians and scientists who tended the world in former epidemics.

When we sit quietly, we can then feel ourselves as part of something greater—part of generations of survivors in the vast web of history and life. We can feel ourselves "being carried," as the Ojibwa elders say, "by great winds across the sky."

This is a time of mystery and uncertainty. Take a breath. The veils of separation are parting, and the reality of interconnection is apparent to everyone on earth. We have needed this pause. Perhaps even needed our isolation, in order to see how much we need each other. Now is the time for love.

As a bodhisattva you can deliberately turn toward the suffering to serve and help those nearby in whatever ways you can. This is the test you have been waiting for. You know what to do.

It's time to renew your vow. Sit quietly again and ask your heart: what is my best intention, my most noble aspiration for this difficult time? Your heart will answer. Let this vow become your North Star. Whenever you feel lost, remember, and it will remind you what matters.

It is time to be the medicine, the uplifting music, the lamp in the darkness. Act. Burst out with love. Be a carrier of hope. If there is a funeral, send them off with a song. Trust your dignity and goodness.

Where others hoard, help. Where others deceive, stand up for truth. Where others are overwhelmed or uncaring, be kind and respectful. When you worry about your parents, your children, your beloveds, let your heart open to share in *everyone's* care for their parents, their children, and their love ones. This is the great heart of compassion. The bodhisattva directs compassion toward everyone: those who suffer, the vulnerable, and those who are the source of this pain. We are all in this together.

As heart wisdom and love matures, we discover we can hold the opposites together. We can carry the trembling beauty together with the ocean of tears that make up human life. We gradually become comfortable with the paradox that is woven into human incarnation: we are made of mud and of stars. We are spiritual beings with a social security number.

We relax into the wisdom of insecurity and love. Life is uncertain. We feel our common humanity—shared longings and fears, love and loss, tenderness and triumphs—and our compassion becomes universal toward all life. We learn to judge less, to let go, to hold things lightly, to forgive and start anew. We know that in the end we can work for the good. We can let go of our fears. We realize that, although we may not be able to perfect our world, we can continue the work of perfecting our love.

As the heart grows to hold the ten thousand joys and sorrows of human incarnation, we are free to respond with intention, rather than reacting blindly. This is the final human freedom: not to be caught in primitive fears but to choose our spirit no matter the circumstances. When Nelson Mandela walked out of Robben Island prison with breathtaking compassion, dignity, and magnanimity, he showed that while your body can be imprisoned, your spirit can be free. This freedom and dignity is available in any circumstance. Dr. Martin Luther King explained, "If a man sweeps streets for a living, he should sweep them as Michelangelo painted, as Beethoven composed, as Shakespeare wrote poetry."

The heart knows we are more than our worst actions. Humans long for love, genuine connection, meaning, and a sense of home. Through heart work, we learn to love more fully and to imbue our days with meaning. We create community and society—not as a false collection of atomized individuals, but as a home we build and tend together. To foster the common good, we respect our differences across all the unique human temperaments and cultures. As a result, our differences are connected to a whole that is now devoted to the common good.

In our hyper-wired world, heart work is an urgent task. Increasingly over the past few years, our "fight, flight, or freeze" response has been deliberately triggered by political discourse and by search algorithms that seek to grab our attention in any way possible. But humans can learn to shift from the primitive brain's fight-or-flight circuit to considered responses from the neocortex and Wisdom from the heart. Buddhist psychology is built on the human capacity to shift from primitive brain reactions and to learn to embody wiser and more compassionate responses.

To expand our circle of compassion is as much our birthright as our own breath. We have the capacities in the human heart and nervous system. Modern neuroscience has shown that human beings are born with innate compassion and care for one's self and others. It also shows that human beings are born with survival circuits that when activated operate from fear, aggression, selfishness, and hate. We have the ability to promote either. As Zen Master Thich Nhat Hanh explains, "The quality of our life depends on the seeds we water. If you plant tomato seeds in your gardens, tomatoes will grow. Just so, if you water the seeds of fear and hate, they will grow. If you water the seeds of peace in your mind, peace will grow. When the seeds of love, respect and peace are watered, we will become happy." We have the capacity to nurture love or to foster hate. Let us choose to nurture love together.

Today, our public discourse seems like it is designed to leave its listeners overwhelmed, disheartened, and feeling powerless. Don't fall for this. We can change the world. We can nurture wellbeing and foster healing. We can value one another. We can find ways to solve international conflicts without war. Compassion directs us to work toward healing in all areas of society.

Our organization, Mobius, is joining with tech companies and institutions to create a code of principles that connect practices of the heart to business practices. Our principles for humane technology represent an effort to adopt the spirit of the Hippocratic Oath within the domain of business and technology. To avoid causing injury and abuse, companies and institutions are asked to agree to the following code of principles:

- We will not create technology that causes harm to humans or life more broadly.

- We will change course and correct our actions, if we learn that we have inadvertently done so.

- We will strive to create technology that fosters human wellbeing and respect.

- We can create technology for profit, but not if it contravenes the first three principles.

- We will act with professionalism in all operations and take these responsibilities as paramount.

This is just one example of compassion guiding our common life together. We are witnesses to the great mystery. Witnesses to truth and love amidst these global crises. We meet the suffering before us and we are knit into a common cloth. If business and technology can adopt empathy, integrity, and wisdom, then our world might come to know justice. The solutions are within you and within me. If we are going to address racism, environmental destruction, warfare, global injustice and now COVID-19, we must engage our compassionate heart in every aspect of our lives. We find ourselves at a pivotal point in history, and the challenges we face must be matched by the inner developments of humanity. These principles hold companies accountable, but we have to connect these practices to *all* spheres of society—to our politics, to our communities, and to our healthcare. We need an educational system based on compassion and understanding. Mindfulness and mutual care help us recall our profound sense of interconnection. This is true cognition. This is the call of our time.

The future of humanity will be found in the heart as much as in the hands and brains of humans. Just as our intention got humans to the moon, the crisis of heart requires an equally powerful Intention—the intention to wed the outer developments of humanity with the inner.

The secret to all this is intention. Like an inner compass, we can set the direction of our life with the deepest intentions of the heart. But we must act well, without attachment to the results. We get to plant seeds based on our best intention, but we do not control when or if they will sprout. Yet with good seeds, as Walt Whitman explains, in their own time, we can expect miracles.

∾

In Zen they say there are only two things. You sit. You tend the garden. You quiet your mind and open your heart. And then, with natural care, you get up and tend the garden of the world.

For you who read this, dear friends, let these words be a reminder, a call, a life mission. Find your way to quiet yourself and tend your heart. Promote love and spread the power of compassion across this mysterious and wondrous earth. Let us together commit ourselves to become a wiser and more loving species.

In spite of the painful current moment, I am anchored in hope. It is time to imagine a new world, to envision sharing our common humanity, to see how we can live in the deepest, most beautiful way possible. Even in this time of difficulty, what we intend and nurture, we can do.

In the end, remember that you are timeless awareness. Recall the consciousness that was born into your body. You were born a child of the spirit. Even now you can return to that awareness. You can become the loving awareness that witnesses yourself reading and feeling and reflecting.

When a baby is born, our first response is love. When a dear one dies, the hand we hold is a gesture of love. Timeless love and awareness is who you are. Trust it. Dear bodhisattva, the world awaits your compassionate heart.

22

Choosing Peace

Francis Mading Deng

WHEN MY UN MANDATE as Special Advisor of the Secretary General on the Prevention of Genocide ended, Secretary General Ban Ki-moon organized a farewell lunch in my honor. As we chatted, I asked him if he was optimistic about the state of the world and quickly added, "Of course, we have to be optimistic; the question is on what basis?" Ban Ki-moon looked up reflectively, then lowered his head, and gave this solemn response: "There has been too much suffering." At first, I thought this was a poor response. But after I reflected on my UN experience, recalled the state of the African continent, and my own country of South Sudan and its devastating civil war; I saw the profound wisdom of the Secretary General's answer. The world is still embroiled in devastating wars that must be stopped.

Over the last two decades I witnessed the worldwide suffering created by these wars as I worked for the United Nations. I served as Representative of the Secretary General on Internally Displaced Persons under Secretaries General Boutros Boutros-Ghali and Kofi Annan, and then served as Special Advisor of the Secretary General for the Prevention of Genocide under Ban Ki-moon. In carrying out these mandates, I discovered that our efforts toward ending this suffering were seen as a threat to national sovereignty. To be effective I had to constructively engage with governments and calm their concerns regarding these threats. I endeavored to recast notions of "sovereignty" in terms of "responsibility." I wanted to demonstrate that my work was in the pursuit of international cooperation and peace. To that end, I applied two principles that have always guided my life's work and experience:

1. Pessimism should be avoided as it leads to a dead end. Optimism—provided it is not blind, but strategically grounded—stimulates creative and productive action.

2. There are almost always opportunities in crises; the challenge is to explore and make effective use of them in seeking constructive solutions.

The end of the Second World War ushered in a new global order guided by the United Nations. This new body promoted self-determination and independence for colonized territories along with respect for human rights and fundamental freedoms. Despite the UN's aspirations, relative peace hung precariously among the major powers. Third-world countries were embroiled in internal wars that, paradoxically, were both managed and aggravated by the Cold War superpowers. These wars massively displaced populations both internally and across international borders and caused untold humanitarian crises. The end of the Cold War raised hopes that a more effective and inclusive global peace and security *could* emerge. This hope was stunted by the rise of international terrorism. Again, Western major powers mobilized a new war effort against international terrorism. But economic globalization continued to divide the world between those who benefitted from it and those it left behind.

Against this backdrop, COVID-19 appeared. It confirmed our global entanglement as infections spread around the world. Paradoxically, localization was affirmed through isolationist responses, and fragmentation continued. Today, upheaval and uncertainty about our world and its future continue. The daunting challenge before us is how to manage this crisis given our local and global contexts.

I believe pursuing peace and responding to humanitarian crises require addressing the root causes of conflicts. In my work at the UN, I wanted to approach these conflicts through the lens of *identity*. Not to magnify differences, but because I wanted to see how the denial of *dignity* and the mismanagement of *diversity* feeds these conflicts. I saw how these two factors classify populations into different groups. Some are offered privileged status with rights and citizenship, while others are denied these rights. Dire consequences arise if we don't respond adequately to such gross injustices. To prevent and resolve these conflicts, we must respect the identity and dignity of every group and promote inclusivity, non-discrimination, and equality. This is a global challenge that affects almost all countries.

During my country missions, I often asked the victim communities what message they wanted their leaders to hear. I got virtually the same response wherever I went. In a Latin American country, the community spokesman said, "That is not our government. To that government, we are not citizens; we are criminals, and our crime is that we are poor." In a Central Asian, former Soviet State, the response was: "That is not our government; none of our people are in that government." In an African country, a Prime Minister said to a senior UN official, "The food you give to those people [his own displaced] is killing my soldiers."

To promote mutual understanding between and within cultures, we have to ensure that equality exists. We must constructively manage diversity, respecting the identity and dignity of every group. In these conversations I encountered a common view: conflict is inherent in human relations, but it is useless to attempt to prevent or resolve it; at best it can only be managed. However, another view—which I hold—considers the normal state of human interaction to be peaceful and cooperative. While the first view normalizes conflict, the second view sees conflict and violence as a negation of the normal state of affairs. In the second view, the aim of conflict resolution is to restore the normal state of relations, which operates on both descriptive and prescriptive levels. If we assume that the normal state of affairs is conflict, then we foster mutual suspicion, apprehension, and confrontation. But if we expect peaceful and cooperative interactions to be the norm, then we encourage more harmonious relations. The difference between these two approaches may be a result of cultural orientation, mindset, or a choice of peace over war.

∾

I have spent much of my adult life studying and documenting the Dinka culture. Although the Dinka are generally proud of their identity and related cultural values, their cultural identity and value system are not widely known. Even among the Dinka themselves, some assume that they know more than they do, and some have been brainwashed by Eurocentric modernization to view their own culture as primitive and outmoded. The Dinka, and indeed other South Sudanese ethnic groups, need to know their cultural heritage in order to engage in a constructive inter-communal dialogue with other groups. This is essential if they are to start choosing peace over war. As they navigate differing fundamental values and institutional structures based on different social orders, it's crucial for them to be familiar with their own cultural identity.

My understanding of Dinka culture is that it begins with the over- riding goal of procreational immortality, which ensures that life continues through successive generations. In my interviews with chiefs and elders about the Dinka worldview, a recurrent theme was that they would speak their minds freely without fear of death—given the politically adverse cli- mate in the country. As one chief put it, "If I die, I have children." Genetic and social immortality through procreation is ensured by a concept known as *Koch e nhom*: "standing the head of the dead person upright." In his in- troduction to my book, *Tradition and Modernization: A Challenge for Law Among the Dinka of the Sudan*, Professor Harold Lasswell termed this "the myth of permanent identity and influence." However, among the Dinka the concept means more than the word "myth" usually connotes. For the Dinka, the concept is more than a fable or story; it aims at ensuring participation and investment in the world around us.

Chief Arol Kachwol of Gok Dinka, a leader whom I interviewed, sum- marized the cultural values of lineage continuity, the dual role of father and mother, and the challenge of social change in the following words:

> It is God who changes the world by giving successive genera- tions their turns. When God comes to change your world, it will be through you and your wife. You will sleep together and bear a child. When that happens, you should know that God has passed on to your children, born by your wife, the things with which you had lived your life. Your father, Deng Majok, if he had lived without a child until his death, his would have been the kind of life that continues only as a tale. But if he left behind a big son who can be spoken of—this is Mading, son of Deng— then, even if a person had never met your father, but he hears that you are the son of Deng Majok in the same way he had heard of your father, he will meet—through you—your father whom he never met.

The value of this concept is not racially or culturally limited to the Dinka but is inter-racially and cross-culturally expandable. Referring to my American wife, Dorothy, who joined me on her first visit to South Sudan, Chief Arol Kachuol introduced the issue of mixed marriages as a bridge across tribal or racial divisions:

> For instance, where is this girl, your wife, from? Is she not from America? And you have brought her back to your country. If you bear a child together now, in that child will combine the

words of her country and the words of your country. It is as though God has given her to Deng Majok, your father. The way we see it, God has brought peace and reconciliation into your hearts. None of you is to hate the race of the other. Tomorrow, the people of that tribe or the people of that race, God will take them and mix them with the people of that race. They too will bear their own races through their children. When that happens, whatever hostility might have been between people should no longer be allowed to continue. Relationship kills those troubles and begins the new way of kinship. Man is one single word with God.

Another key principle of the Dinka value system is *cieng*, a concept of relationships aimed at unity, harmony, and conciliatory management of differences. The anthropologist Godfrey Lienhardt, who focused his study on the Dinka, and Father Arthur Nobel, the pioneering Catholic missionary who extensively studied the Rek Dinka, both elaborately discussed *cieng* as representing the Dinka notion of how society should be ordered. The concept has a wide range of interrelated meanings, including custom, law, behavior, conduct, and way of life. *Cieng* is both prescriptive (what ought to be) and descriptive (what actually happens). The concept is sometimes referred to as *cieng e baai*, and *baai* means home, village, community, tribe, or country. *Cieng* is therefore specific to a social unit and includes expanding circles, ultimately embracing humanity.

The Dinka believe that *cieng* should include respect not only for all humans as God's creatures, whatever their race, culture, or religion, but also non-human creatures of God. It requires being in harmony with humanity and nature. As Chief Thon Wai put it to me:

> If you see a man walking on his two legs, do not despise him: he is a human being. Bring him close to you and treat him like a human being. That is how you will secure your own life. But if you push him onto the ground and do not give him what he needs, things will spoil [for you] and even your big share, which you guard with care, will be destroyed. Even the tree which cannot speak has the nature of a human being; it is a human being to God, the person who created it. Do not despise it. It is the government which has taken harmony away.

In the words of Chief Makuei Bilkuei, from another Dinka tribe:

> It is the government which has taken harmony away. We were one with our hyenas, with our leopards, with our elephants, with our buffaloes . . . We should all combine—the people, the

animals, the birds that fly, we are all one. Let us all unite . . . Even the animals that eat people, let's embrace them all and be one.

Another fundamental concept among the Dinka is *dheeng*, which can best be translated as dignity. Closely associated with *dheeng* is *atheek*, respect, which is central in human relationships. *Dheeng* and the related value of *atheek* include physical appearance, the aesthetics of beauty, artistic expression in song and dance, and the recognitions associated with them. Like *cieng*, these values incorporate proper conduct in relationship to others

These social and moral values also apply to leadership. When a person assumes authoritative control of the family, clan, community, or tribe he is said to *dom baai*, and *baai*. As I said, this concept encompasses home and expands to include community, tribe, and country. *Dom* also applies to physical possession or control, but it's not just physical control. It also carries with it resonances of pacification, ensuring peace, security, and order. The next requirement is *guier* (putting in order) *baai*, which means reforming and improving the situation by solving any problems that existed before assuming control.

Two other concepts relate to physical control and the exercise of authority in introducing reforms and ensuring stability. One is *mac baai,* which internally means to tie or bind, a term normally used to mean tying a cow to a peg with a rope. *Mac* also refers to family or kin. The other word is *muk baai*, which means "keeping," a word that also applies to cattle and to nurturing a child and that implies stability. These cultural values are mutually reinforcing. To have *mac* or *muk baai* is also to have *dom baai,* and this gives shape to the normative meaning of authoritative control. To be a successful leader, one has to live up to these values. All these principles and the values dovetail and demonstrate how benefits are carried with responsibilities.

There are other concepts that are expressed in words that relate to leadership and the administration of justice. One word is *yich*, which literally means "truth," but also means "right." The expression *Ke yich* simply means "It is the truth." But to say *ke yich du* means "It is your truth," which means "It is your right." The other word is *luk*, which means resolving a dispute. But it also means to persuade a person to a point of view. The combination reflects the Dinka normative process of settling disputes, which is not an adversarial determination of rights and wrongs, as is the case in Western judicial processes, but a conciliatory process of resolving a dispute in an amicable way aimed at reaching a consensus and achieving reconciliation.

The chief among the Dinka does not wield coercive force and impose his will on his "subjects"; he is first a peer among equals. The chief is a moderator whose authority and effectiveness rests on his managing relations between and among his people through his ability to persuade. His only coercive function is grounded in social affirmation for compliance with his decisions and condemnation, ostracism, or, in extreme cases, divine punishment for disobeying him. The Dinka normative process of settling disputes is not adversarial in its determination of rights and wrongs. In contrast to the Western judicial process, it is a conciliatory activity that resolves disputes in an amicable way, reaches a consensus, and achieves reconciliation.

~

The Dinka value system as reflected in my various books resonates with non-Dinka readers nationally, regionally, and internationally. The normative principles of *cieng* are almost identical to the world-famous concept of *ubuntu*, which Nelson Mandela, Archbishop Desmond Tutu, Thabo Mbeki, and other African leaders and scholars have helped to universalize. Essentially, *ubuntu* is a concept of shared humanity in which the interest of the individual is in harmony with the community or humanity.

Taking my country of South Sudan as a starting place, I believe that these cultural principles form a holistic model and are relevant for developing a culturally oriented normative framework of good governance, conflict prevention, and nation building. We need to have a better appreciation of our own local cultures if we are to foster constructive dialogue and promote mutual understanding, accommodation, and cross-cultural processes of equitable co-existence or integration.

This process of dialogue within, between, and among cultures can then be extended to more inclusive regional and international contexts. In his foreword to my book, *Identity, Diversity, and Constitutionalism in Africa*, General Olusegun Obasanjo, former president of Nigeria, wrote: "I share Dr. Deng's view of an Africa that builds on its time-tested cultural ideals and institutionalized practices . . . I might also note that these values have much to offer not only Africa but the world as a whole. Just as Western democracy enshrines certain universal values, so does the African worldview."[1]

A policy-oriented, normative way of framing the challenge of constructively managing diversity should be grounded in four interconnected principles that apply in many (if not all) cross-cultural relations and that are related to the four interconnected levels of transition toward an inclusive framework of values. These principles are *identity, dignity, diversity, and equality*. The correlative levels are *local, national, regional, and global*.

---

1. Deng et al., *Identity, xii.*

Identity is a fundamental characteristic of every individual and group. It is the core of our shared universal humanity. Dignity, however defined, is a universal value of humanity; this is why human rights norms are grounded in human dignity. Diversity at the point of contact raises the issue of comparison and discrimination. Equality becomes the inevitable demand against unacceptable stratification. Since each group in isolation views itself subjectively as the ideal model of God's creation, they are all inherently equal in their mistaken assumptions. The myth that all men and women are born equal betrays what ought to be as if it already existed—a merger of identity, dignity, diversity, and equality.

∾

I would like to end with the two principles with which I began: strategic optimism, and seeking opportunities in crises. COVID-19 has imposed on the world a paradoxical process of infectious interconnectedness that reaffirms globalization while also generating an isolationist, distancing response more conducive to localization. As we develop a better understanding of the virus and its impact across social, economic, and political fields, a silver lining may be that nations and communities will be forced to seek solutions by resorting to the basic elements of internal resilience, creative self-reliance, and resourcefulness out of necessity. This is the challenge facing the newest country of the world, South Sudan.

South Sudan's independence movement was a liberation struggle against Arab-Islamic domination; it aimed at creating a New Sudan of full equality without discrimination on the bases of race, ethnicity, religion, gender, or culture. Dr. John Garang de Mabior was the founding leader of the Sudan People's Liberation Movement, which championed the struggle. Dr. Garang de Mabior stipulated several policy objectives: invest the accruing oil revenues in fueling the engine of agriculture, take towns to the people in the rural areas, and build "roads, roads, roads." His sudden death in a plane crash turned that vision into an unfulfilled dream. The urgent need of COVID-19 now presents us with the opportunity to revive his vision. This is not just a vision for South Sudan, where the consequences of putting it on hold are all too apparent; it's the situation we face all across the world.

The revival of Indigenous resourcefulness should lead to the restoration of pride and dignity. It should foster a regeneration of the self-confidence to engage in a constructive dialogue and promote mutual understanding. It should sow the seeds that enrich cross-fertilization domestically, regionally, and globally. This is of course a daunting task, yet one that must be optimistically and creatively pursued to reconnect localization and globalization. The constructive management of diversity remains the overriding strategy for promoting the principles of equitable peace and cooperation between and among races, ethnicities, religions, and cultures. COVID-19 must be met with creative and strategic optimism as we explore the possibilities and respond with constructive solutions.

References

Deng, Francis Mading, Daniel J. Jiménez, David K. Deng, and Vanessa Jiménez. *Identity, Diversity, and Constitutionalism in Africa.* Washington, DC: Institute of Peace, 2008.

23

Integral Ecology

FR. JOSHTROM ISAAC KUREETHADAM

The Moral Compass of Integral Ecology

ONE OF THE MOST poignant moments of the lockdown during COVID-19 happened on the evening of 27 March 2020. As dusk fell on the sprawling St. Peter's Square, there was an eerie silence, broken only occasionally by the shrill sirens of speeding ambulances. A pensive man clad in white walked alone in the drizzling rain toward the podium set up in front of the grandiose façade of the Basilica. It was Pope Francis. In Italy alone, more than a quarter of the national population watched him on television, and millions more joined them around the world. What followed was a special and intense moment of prayer for the numerous victims of the invisible coronavirus that was galloping across Italy and the rest of the world.

Moments later, Pope Francis lifted the Blessed Sacrament and delivered the *Urbi et Orbi* blessing. Like the two baroque arms of Bernini's colonnades that embrace the Square, Pope Francis too was embracing the world—especially the sick and the dying—and all those who cared for them.

It was an incarnation of the loving embrace of integral ecology. Pope Francis had introduced this concept five years ago in his landmark encyclical *Laudato Si'*, in which he invited all people of good will to come together to care for our common home.[1] If Earth is our common home, we are all a common family, and as in every family, in moments of grief and trial, we must embrace the most vulnerable among us.

The perspective of "integral ecology" offered in Pope Francis' encyclical letter *Laudato Si'* throws open a promising window of opportunity in our response to the current COVID-19 emergency. The encyclical's core intuition that everything is interrelated and interdependent provides a framework to understand, analyze, and respond to the coronavirus. *Laudato Si'* reminds us that we need to think and act from an integrated and planetary perspective. If we are to overcome our current crisis, along with the host of related physical and socio-economic crises that humanity is facing, we need an inclusive approach.

The integral ecological perspective of *Laudato Si'* requires that we attempt a comprehensive understanding of COVID-19. We also need to work toward an accurate diagnosis of its deeper causes and offer an integrated collective response. Through this work, we can sketch the contours of a more sustainable and harmonious relationship between humanity and the natural world. I will follow Pope Francis' moral and spiritual guidance in *Laudato Si'* to suggest a way forward on this journey.

An Integral Understanding of the COVID-19 Emergency and Related Crises

If "everything in the world is connected," no single problem can be considered in isolation. There are no silver bullets for the challenges we face, much less for the crisis created by COVID-19. We cannot make the mistake of treating this virus as a mere health crisis that will be solved with an effective vaccine. Nor can we dupe ourselves into thinking that the social impacts can simply be addressed with massive economic stimulus packages. Instead, we will need to recognize the complexity of the present crisis, the multiple impacts of which are already spilling over into the realms of social life, economics, trade, politics, and inequality.

The current coronavirus crisis must be understood in its complexity and entanglement with other crises and then addressed with "epochal challenges." In fact, the COVID-19 emergency is only the most recent of a series of warnings indicating that we are indeed approaching crucial geophysical and socio-economic tipping points.

---

1. See Francis, *Laudato Si'*. All further citations from this encyclical will be footnoted with the paragraph number indicated.

We must begin to see the coronavirus crisis within the wider context of humanity's increasingly antagonistic relationship with the natural world. The scientific community has been reminding us for several decades now of this larger view and the consequential nature of these changes. COVID-19 presents a clear warning that we have transgressed crucial physical tipping points in our relationship with the natural world.

We can recall the monumental studies of "planetary boundaries" in 2009 and 2015, according to which we have already crossed crucial thresholds in at least four areas: climate, biosphere integrity, land system change, and biogeochemical flows (due mainly to modern agriculture). Of all these, the climate and biodiversity crises are the most critical. According to these indicators, Earth is dangerously poised on the verge of a sixth mass extinction of species, with disastrous implications for the entire web of life. Our climate indicators fare no better, as the 2018 report from the Intergovernmental Panel on Climate Change made abundantly clear. Humanity has just twelve years to ensure that global average temperature does not cross the critical threshold of 1.5° C if we are to avoid immensely catastrophic consequences.

The coronavirus has to be situated within the wider context of the collapsing planetary boundaries that directly threaten human welfare—the very survival of all human civilization as we know it. The COVID-19 crisis is part and parcel of recent warnings from the natural world. For example, the Amazon and Australian wildfires, the unprecedented floods in East Africa followed by the locust invasion of biblical proportions, the unprecedented melting of the Arctic and the Western Antarctic, and the most recent finding that the European winter was 3.5° C above normal. When addressing the effects of COVID-19, Pope Francis called attention to our collective *amnesia* (forgetfulness) surrounding these messages the natural world is sending us:

> Who now speaks of the fires in Australia, or remembers that eighteen months ago a boat could cross the North Pole because the glaciers had all melted? Who speaks now of the floods? I don't know if these are the revenge of nature, but they are certainly nature's responses.

If we want to understand the gravity of this global pandemic, we also need to grapple with its tragic and far-reaching socio-economic impacts. We need to listen to the "cry of the Earth and the cry of the poor."[2] "Let anyone with ears listen!"[3] COVID-19 has brought about the worst global crisis since

2. Francis, *Laudato Si'*, 49.
3. Matt 11:15.

the Second World War. It is estimated that the economic fallout from this global pandemic could increase global poverty by half a billion people. This is eight percent of the human population. It would literally mean reversing a decade of global progress on poverty reduction, and we could see global poverty increase for the first time since 1990. This poses a significant challenge to the UN Sustainable Development Goal of ending poverty by 2030.

All of humanity will be affected by the current coronavirus emergency, but it will hit the poor hardest:

> This virus affects us all, even princes and film stars. But the equality ends there. By exploiting the extreme inequalities between rich and poor people, rich and poor nations, and between women and men, unchecked this crisis will cause immense suffering.[4]

This pandemic will only further exacerbate the unsustainable practices of our divided world. Before the coronavirus, we were living in a world of extreme and growing inequalities. We live day to day as a divided family—821 million members of our common household cannot afford a single meal a day. Millions lack primary healthcare facilities. Poverty and environmental degradation have caused many to flee their homes. These colossal challenges abound, while the mad and reckless arms race and nuclear proliferation cost national economies trillions of dollars. This money should have been spent to ensure basic amenities like food, healthcare, decent housing, and basic education for children. Our efforts should be directed toward caring for those with the greatest need in our common household. We live in an increasingly global, interdependent, and environmentally constrained world, and COVID-19 has provided a prism through which to see these challenges more clearly.

The Multiple Roots of the COVID-19 Emergency

The integral ecological approach invites us to see the multiple roots of the current COVID-19 emergency as we respond. To address these root causes we need to adopt an ecological perspective attuned to biodiversity. The cure for any malaise can only begin with an effective diagnosis of the underlying causes, and our shared climate crisis increasingly contributes to the emergence and spread of infectious pathogens like COVID-19.

Although terrible pandemics are woven throughout the fabric of human history over millennia, in recent years we have witnessed an

4. Oxfam, "Dignity Not Destituion."

unprecedented rise in emerging infectious diseases. In just the past few decades, three hundred or so new infectious pathogens have emerged as habitats are destroyed—violating the integrity of ecosystems. It is important to remember that the novel coronavirus, like its predecessors, is a *zoonotic* disease: a microbiologic infection acquired from animals, passed to humans, that leads to human-to-human spread. As natural habitats are destroyed and people move into closer contact with animal species that carry deadly viruses, pathogens can jump to humans and lead to explosive human-to-human transmission. Most of the emerging zoonotic infections (nearly seventy percent) are linked to human environmental changes.

The origin of COVID-19 and many others—like SARS, MERS, and, analogously, the Ebola outbreak—has to do with human interference in the intricate balance of natural ecosystems. This occurs through wildlife trading; deforestation through mining, logging, and animal husbandry; and agriculture. These activities destroy local biodiversity and clearly contribute to the root causes of the increased emergences of new viruses in recent decades. Human activity is placing unprecedented pressure on the natural world, and we are now experiencing the unbearable consequences.

COVID-19 reveals a fundamental truth that we have ignored for too long: we cannot be healthy unless our relationship with the planet and its ecosystems is healthy. Human health and planetary health are intimately linked together. We need to realize and accept that we "are not completely autonomous . . . We cannot interfere in one area of the ecosystem without paying due attention to the consequences of such interference in other areas."[5]

With the rapid destruction of Earth's life-sustaining ecosystems, we increase the risk of new and possibly deadlier human-adapted viruses in the future. A clarion call is ringing around the world. We have to realize that we cannot flourish, or even exist, if we destroy the very ecosystems that sustain the whole fabric of life. Nature is sending us a clear message: if we ignore the clear linkage between human, wildlife, and environmental health, we do so at our own peril.

Climate change threatens to disrupt the basic life-support systems that ultimately underpin human health and wellbeing. Many of the infectious disease agents such as protozoa, bacteria, and viruses are highly sensitive to climatic conditions like temperature and rainfall. Any changes in climatic patterns and seasonal conditions affect the spread pattern and diffusion range of zoonotic pathogens. This means our impact on these environments could increase the emergence of dangerous pathogens.

5. Francis, *Laudato Si'*, 105, 131.

But the integral ecological approach invites us to go a step further and plumb for deeper roots. Why do we relate to the natural world in this way? What compels the decimation of biodiversity, accelerates the climate crisis, and adds to pollution, waste, and the unsustainable depletion of our natural resources? At the base of these vast crises is a worldview, a mindset, that promotes aggressive, domineering, and destructive relationships with the natural world. The deeper roots of our global crises lie in the current technocratic and economic paradigm which itself is based on "excessive anthropocentrism" and "rampant individualism."[6] These are postures that continue with no appearance of slowing down. They occur to the detriment of the natural world and the poor and vulnerable members of our common household. This is what Pope Francis calls in *Laudato Si'* the "dominant technocratic paradigm."[7] He reminds us:

> A certain way of understanding human life and activity has gone awry, to the serious detriment of the world around us . . . I propose that we focus on the dominant technocratic paradigm and the place of human beings and of human action in the world . . . [T]he deepest roots of our present failures . . . have to do with the direction, goals, meaning and social implications of technological and economic growth.[8]

An Integral Response to the Covid-19 Emergency: Solidarity and Care

COVID-19 has caused major disruptions to health systems, economies, and societies, but we have failed in offering an effective global response. In the face of this serious pandemic, we have lacked an internationally co-ordinated approach. Our response to the multiple crises compounded by this global public health emergency has been anything but united. Precisely when we should be joining hands in solidarity to fight a common enemy, some of our prominent leaders have preferred parochial and nationalistic sentiments and indulged in petty blame games. However, the coronavirus does not respect national boundaries or align with political agendas, and we have failed to stem the tide as new cases and deaths continue to rise in many parts of the world.

6. Francis, *Laudato Si'*, 116, 162.

7. Francis, *Laudato Si'*, 101.

8. Francis, *Laudato Si'*, 101, 109.

The sad state of our current affairs was driven home by none other than the World Health Organization's Director General, Dr. Tedros Ghebreyesus: "The greatest threat we now face is not the virus itself. Rather, it is the lack of leadership and solidarity at the global and national levels . . . we cannot defeat this pandemic as a divided world."[9]

How do we move out of this stalemate? What helpful insights can come from integral ecology in responding to this unprecedented emergency?

Pope Francis almost anticipated this question in his address to the Special General Audience on the fiftieth anniversary of Earth Day. He said, "As the tragic coronavirus pandemic has taught us, we can overcome global challenges only by showing solidarity with one another and embracing the most vulnerable in our midst." Our response must be guided by the core human values of solidarity and compassion.

Our common identity as members of the one human family is deeper than all our geographic, political, religious, and cultural diversities. We need to heal our fragmented thinking embedded in how we view each other and our world. "Interdependence obliges us to think of *one world with a common plan*."[10]

To grow in solidarity, we need to think and act in an integral way. As "everything is closely interrelated and today's problems call for a vision capable of taking into account every aspect of the global crisis," we will need to employ an integrated approach in responding to the multitude of crises plaguing us.[11] As Pope Francis writes, "It is essential to seek comprehensive solutions . . . Strategies for a solution demand an integrated approach to combating poverty, restoring dignity to the excluded, and at the same time protecting nature."[12] Multilateralism is a core constitutive element of an integrated strategy: "A global consensus is essential for confronting the deeper problems, which cannot be resolved by unilateral actions on the part of individual countries."[13]

In the face of our fragmented world, we are invited to grow in our compassion toward one another and the natural world. Compassion is best expressed through care. We have witnessed an outpouring of compassionate care in the wake of the coronavirus explosion. We have seen it in the sacrifice and dedication of thousands of healthcare workers: doctors, nurses, and all other medical staff. We have observed the selfless commitment of fellow

9. Chappell, "Lack of Unity."

10. Francis, *Laudato Si'*, 164.

11. Francis, *Laudato Si'*, 137.

12. Francis, *Laudato Si'*, 139.

13. Francis, *Laudato Si'*, 164.

citizens who have toiled to guarantee basic services: volunteers, supermarket employees, cleaners, caregivers, providers of transport, law and order forces, etc.

To care is an expression of our humanity. An ethics of care is rooted in a specific view of the world as a network of relationships. In this time, "we must regain the conviction that we need one another, that we have a shared responsibility for others and the world."[14] Special consideration must be given to our most vulnerable brothers and sisters, just as we would offer in our natural families. We need to answer the call to embrace, with particular affection and care, those who are suffering most in our global household. Caring is also what makes us more God-like. When we care, we are indeed imitating God's loving, tender care toward all creatures. This is our original human vocation. As Pope Francis writes in *Laudato Si'*, we are called to "cooperate as instruments of God for the care of creation, each according to his or her own culture, experience, involvements and talents."[15]

Birth Pangs of a New World?

Every crisis is an opportunity, as the Greek etymological origin of the term reveals. The experience of our collective fragility could inspire us to build together a better world. The COVID-19 emergency can indeed be the prelude to a new beginning—a new dawn for humanity and for our common home. May we leave behind the unabashed personal egoism and unfettered collective market greed to the detriment of the poor and the planet. May we be guided by an integral ecology that sees the interdependence of all of life. May our present groaning and toil be the birth pangs of a new way of living together, bonded in love, compassion, and solidarity, and a more harmonious relationship with the natural world. May our collective and compassionate response to the coronavirus emergency usher in a more inclusive, peaceful, and sustainable world.

References

Chappell, Bill. "Lack of Unity Is a Bigger Threat than Coronavirus, WHO Chief Says in Emotional Speech." *NPR*, July 9, 2020. https://www.npr.org/sections/coronavirus-live-updates/2020/07/09/889411047/lack-of-unity-is-a-bigger-threat-than-coronavirus-who-chief-says-in-emotional-sp.

14. Francis, *Laudato Si'*, 229.
15. Francis, *Laudato Si'*, 14.

Coogan, Michael, Marc Brettler, Carol Newsom, and Pheme Perkins. *The New Oxford Annotated Apocrypha*. 5th ed. New York: Oxford University Press, 2018.

Oxfam. "Dignity not Destituion."*Oxfam International*, April 9, 2020. https://www.oxfam.org/en/research/dignity-not-destitution.

Francis, Pope. *Laudato Si'*. The Vatican, May 24, 2015. http://www.vatican.va/content/francesco/en/encyclicals/documents/papa-francesco_20150524_enciclica-laudato-si.html.

COMMUNITY

Somos Uno
by Lavie Raven

24

Coming Back to Place

HELENA NORBERG-HODGE

As THIS BOOK CONFIRMS, the COVID-19 pandemic offers numerous powerful lessons. Through poignant images of empty supermarket shelves and the worldwide scramble for surgical masks, it reveals the intrinsic fragility of our global supply chains stretching across the world and back again. Even the mainstream media argues that governments will no longer want to remain so dependent on international trade for critical needs.

As you might have guessed, the local food movement was greatly bolstered by the pandemic. Small farmers met with unprecedented demand, restaurants became producer-consumer hubs, and people planted vegetable gardens in numbers not seen since the "victory gardens" of World War II.[1]

Other forms of local self-reliance emerged. When big, centralized institutions failed to protect the most vulnerable, self-organized support groups arose to meet needs. These groups coordinated grocery and medicine deliveries and helped provide for those whose jobs vanished.[2] At the same time, neighborhoods were enlivened by once-atomized residents singing to each other from balconies, open windows, and across streets.

---

1. Local Futures, "Let's Localize."

2. As examples, see the following organizations: https://covidmutualaid.org/; https://www.usacovidmutualaid.org; and https://viralkindness.org.au.

Perhaps most importantly, the international response to COVID-19 has demonstrated the potential for rapid, large-scale change. But, in the wake of the shock, what form will such change take? Our multiple crises—climate change, species extinction, the erosion of democracy, an epidemic of depression, the rise of right-wing authoritarianism, and many more—tell us that we must shift away from the volatile, resource-intensive, job-destroying and inequitable global economy. We have to move towards decentralized, community-based, local economies based on respect for Nature—both human and wild.

But is the global pandemic enough of a wake-up call? Will it encourage us to build political movements that ensure systemic support for a localizing path? Or will it serve primarily as an excuse to rush even faster toward a high-tech dystopia—toward a fully globalized, corporatized economy built on algorithms, robots, and constant surveillance? This is the historic fork in the road that COVID-19 marks.

There is reason to hope that we will, collectively, move in a saner and healthier direction. A groundswell of grassroots support for localization has been growing for decades, as increasing numbers of people see through the shallow façade of consumer capitalism. We are recognizing that *connection*, both to others and to Nature, is the true wellspring of happiness.

We are also discovering that smaller-scale, slower-paced economic systems are conducive to restoring ecological health and creating more abundant, secure livelihoods. Localization offers a systemic solution to many of the modern world's most pressing problems. So it isn't surprising that, in almost every country, people are starting farmers' markets, community co-ops, local business alliances, regenerative agriculture projects, place-based education schemes, nature connection therapies, and a myriad of other initiatives that reweave relationships at the local level.

These trends are inspiring testaments to human ingenuity and goodwill. For that reason, my heart breaks when I hear things like "humanity is a cancer on the Earth, and we deserve to extinguish ourselves." Words like these are now a common refrain within environmental circles. But the belief that human beings can only be a force for destruction is a product of a distorted view of human nature—one that sees us as innately greedy, selfish, competitive, and violent. These are not innate features of human nature; they are the product of an economic system that transforms us from cooperative and caring members of community into isolated, self-interested consumers. That system threatens to change us from *homo sapiens* into *homo economicus* (the economic animal).

～

Tragically, most of our political and business leaders remain locked in a globalizing formula for "economic growth" that depends on treating us as *homo economicus*. The recipe also demands constant escalation of resource use, ever more commodification (putting a dollar value on neighborly help and every aspect of the natural world), and the continuous expansion of international trade. For decades, our leaders have poured taxpayer money into creating vast technological, energy-intensive systems for centralized profit-making. From satellite communications to artificial intelligence to the "internet of things," these systems make human skill and knowledge redundant. They continue to expand the infrastructure for international trade, fund the biggest militaries in human history, and systematically deregulate global institutions. These practices have occurred to such an extent that multinational corporations and financial institutions are now beyond the reach of democratic control.

The sheer scale of this techno-economic system enforces standardization and unecological production at a structural level. The system enforces cultural homogenization through top-down, authoritarian control. Thanks to "free trade" treaties, monopolistic global agribusinesses like Unilever, Nestle, and PepsiCo are able to market vast quantities of uniform products in countries around the world. They create market conditions that favor factory farms, vast chemical-intensive monocultures, and equally monopolistic supermarket chains. It is in the economic interest of global companies like these to erase cultural diversity: homogenized tastes and buying habits make corporate advertising and marketing more "efficient" and profitable.

Meanwhile, the same "free trade" treaties give global manufacturing businesses the ability to move wherever labor costs, taxes, and environmental standards are lowest. This makes jobs insecure everywhere; stable, cohesive communities are increasingly made vulnerable.

On so many levels—economic, social, environmental—the globalized economic model is utterly destructive. So why do policymakers continue to promote it?

I believe that, more than anything else, it is because of blindness. In their high-rise office buildings in Washington, DC and Brussels, policymakers and corporate advisors are largely insulated from the real-world impacts of their policies. From their gated communities it's possible to retain faith in the neoliberal fairytales of "trickle-down" economics, "comparative advantage," and "progress" through GDP growth. Policymakers can remain ignorant of the accelerating destruction of Nature and the intensifying competition for dwindling jobs that is the lived reality for hundreds of millions of people.

So, they strip away democratic oversight on the activities of multinational corporations. They write into law additional tax breaks and provide still more subsidies. Without full awareness of what they are doing, they are reengineering society into ever greater dependence on distant, unaccountable entities. They are subsidizing centralization and ecological destruction—they are breeding social unrest.

Earlier this year, Tim Leunig, a high-level advisor to the UK treasury, issued a report claiming that agriculture only contributes about one percent to the nation's GDP. "Let's be like Singapore and import all our food," he reasoned, "let's completely stop farming in the UK." This kind of reductionist thinking brings with it an incredible hubris: he may as well have added, "let's throw all our farmers out of work and have unseen slaves on the other side of the world grow all our food for us."

Mr. Leunig and others like him express the layers upon layers of ignorance that construct globalist economic thinking: ignorance of the huge transport miles it entails, and of the fossil fuels and resource-extraction that it depends on. Ignorance of the profound ways in which a centralized food system promotes profoundly unethical, unhealthy, and unsustainable modes of production. Ignorance of the criminal wastage of food in such systems, and to the slavery-like conditions of the producers. Ignorance of the fact that the human body needs food as fresh as possible for healthy functioning.

Unless we challenge such GDP-obsessed "logic," we will be taken—step by step—into the techno-globalist's dream of the future. It is a dream of total urbanization and dominion over Nature. A dream in which wild spaces, community, cultural diversity and genuine democracy have no place. According to Google's Ray Kurzweil, our food will come from "AI-controlled vertical buildings" and include "in-vitro cloned meat." Tesla's Elon Musk informs us that building a city on Mars is "the critical thing for maximizing the life of humanity," while "thirty layers of tunnels" will relieve congestion in Earth's high-density cities. Goldman Sachs explains that the digitization of everyday objects will "establish networks between machines, humans, and the internet, leading to the creation of new ecosystems that enable higher productivity, better energy efficiency, and higher profitability."

These ideas are lauded as visionary and bold, but what they promise is simply the escalation of dominant trends—neo-colonial expansion, urbanization, and commodification—turbocharged with fancy gadgets. Even much of the environmental movement has been co-opted by this vision. A well-meaning "Green New Deal" wants to plaster fertile soil with solar panels and mountaintops with industrial windmills in order to fuel the economy.

Behind the facades of "green technology," the techno-globalist future is built on rare earth minerals trawled from the seabed, and propped up by the exploitation of tens of millions of anonymous workers in data warehouses, synthetic food-processing plants, cobalt and lithium mines, electronic sweatshops, and prisons.

In recent years, everything from climate change to the rise of China has been used as excuses for increasing public investment in the infrastructure for this techno-economic expansion. Now, the shock created by COVID-19 is a windfall for the same corporate interests who are driving an unprecedented leap into the digital, "no-touch" future. In the name of public health, democratic checks and balances are being skirted to give the green light to unprecedented systems of surveillance and information mining. Any meaningful conversation about, for example, the effects of digitization on employment, or the consequences of remote learning on children's psychosocial development, are completely bypassed.

We need to zoom out, examine the big picture, and interrogate this vision, which is broadcast to us in everything from TED talks to Netflix series:

- If increases in screen-time and social media usage correlate with jumps in depression and addiction rates among young people, what will a remote world do to our wellbeing?[3]

- If companies like Amazon, Airbnb, and Uber are already evading taxes, destroying small businesses, and replacing secure livelihoods with low-paid, unprotected jobs in the "gig" economy, what toll will technologization take on job security? Will robots replace human beings in every sphere—from farming, to care for the elderly, to education and architecture?[4]

- If Facebook and Google are already selling our data to those who want to tilt elections in their favor, what will become of our democracies when even more of our personal and societal information is commodified and privatized?[5]

3. Boers et al., "Association of Screen Time."
4. Gardner, "Amazon in Its Prime"; Opinion Has It, "Our Digital, No-Touch Future."
5. Pariser, "Beware Online 'Filter Bubbles.'"

- If reductionist thinking linked to unfettered markets got us into this mess in the first place, why should we believe that even more of the same will offer genuine solutions to our ecological crisis?

∼

Okay, enough of the scary stuff. Let me reiterate that the techno-globalist path is not the only path illuminated by COVID-19. There is another vision for the future, emerging at the grassroots on every continent. As people yearn for the deep bonds of community and connection to Nature, they are pushing—from the bottom up—for a fundamental shift in direction. They are pushing towards localization. And as the specter of a crisis-ridden future sets in, more and more of us are joining this movement.

It is not a vision built upon a few billionaires' fetish for high-tech gimmicks and knacks for money-accumulation. Instead, it emerges from a deep experience of what it means to be human. We don't wish to transcend the confines of the natural world, but to prosper in harmonious balance *with* the biosphere. We don't want technology to eliminate all human labor, but rather to engage in meaningful, productive, and secure livelihoods. We don't strive for material measures of competitive "success," instead we aspire to be part of vibrant, diverse, intergenerational, and intimate communities.

Through thousands of grassroots initiatives, people are coming together to reweave the social fabric and to reconnect with the Earth and her ecosystems. My organization and I are in the fortunate—and all-too-rare—position of receiving news of new localization projects every week. These projects are springing up in all corners of the globe. Stepping back and looking at the bigger picture can leave us amazed at what people are accomplishing. Given the huge systemic supports for the big and global economic and political systems, the continued flourishing of these alternatives is a testimony to the power of community. It is a witness to the motivation, perseverance, and strength that emerges when people come together to create positive change.

These smaller movements and communities are also connecting with one another across borders, to form larger "new economy" networks and international alliances. One example is La Via Campesina—the world's largest social movement, which represents 200 million small farmers worldwide. These alliances push for policies that support local economic sovereignty based upon smaller producers and localized businesses. This is part-and-parcel with their demand to level the economic playing field by re-regulating multinational businesses and banks.

This emerging movement transcends the polarized left-right dichotomy. Far from promoting big government or big business, it's about enabling diverse human values and dreams to flourish, while simultaneously re-embedding culture in Nature. It's about enabling societies to withdraw dependence on undemocratic, monopolistic, resource-intensive entities that exploit people on the other side of the world. Instead, it's about supporting local production that prioritizes local needs. The emphasis here is on *real* needs, not the artificial wants created by marketers and advertisers in an effort to stoke the furnaces of consumerism and endless growth.

Ultimately, the wide range of localization initiatives that make up this emerging movement reflect an enduring and innately human desire for love and connection. They emphatically demonstrate that human nature is not the problem. On the contrary, the problem is the *in*human scale of a techno-economic monoculture that is fragmenting us from each other, manipulating our dreams, and reducing our humanity.

This understanding is reinforced by observing what happens when people come back into contact with human-scale structures. Through localization, we rediscover joy and contentment. We build more secure identities. In projects that involve therapeutic horticulture, animal connection, and community gardening, I have seen depression healed and delinquent teenagers and prisoners transformed—I have seen people given purpose in their lives.

As we rebuild the fabric of local interdependence, we align with our internal natures. We align with our compassion, our joy, our empathy, our sense of beauty, and our indelible desire to connect. This not only heals us, but it heals entire ecosystems. When we depend on the land around us, it makes no sense to plunder it for short-term gain. Instead, we are inclined to use our hands, our hearts, and our intelligence to nurture that land. This posture increases its productivity and nutritive capacity and tends to the health of the larger ecosystem.

On an overcrowded planet, it therefore makes ecological *and* economic sense to localize as much as possible. We must put our intelligence to work in restoring the local ecosystems and community relationships on which we all ultimately depend.

Shorter distances mean more eyes per acre. This translates into substantially more jobs and more innovative uses of available resources. At the same time, as we withdraw dependence on highly centralized, automated systems like healthcare and education, we can rebalance the ratios between doctor and patient, between teacher and student, and thereby make space for individual needs and capabilities.

Unemployment, resource wastage, poverty, and biodiversity loss are all political decisions that, at the moment, are being made according to the mantra of "efficiency" in centralized profit-making. It is not at all utopian to imagine a world without those problems.

It is also far easier to start creating this world than to continue flogging the dead horse of corporate globalization. Localizing wouldn't entail sending fleets into the middle of the Pacific Ocean to vacuum up rare minerals from its dark depths, nor plastering the globe with sensors to monitor carbon emissions. It wouldn't necessitate the construction of mega-cities and the vast infrastructure required to serve a one hundred percent urbanized population. Nor would it lead to the wars, climate catastrophes, pandemics, and migration crises that will inevitably disrupt the lives of everyone if we stay on the current path.

Instead, embarking on the path of localization means allowing the rural half of humanity to avoid being pushed or pulled into cities through economic and psychological forces. It means connecting cities to regional producers of food, clothing, and building materials wherever possible. It means starting a gradual process of de-urbanization to repopulate the rural towns and smaller cities that have been abandoned by the global economy. It means conserving what is left of the natural world. We can help Gaia restore her ecosystem processes—no amount of carbon-sucking machines or "precision" pollination technology will ever provide an adequate substitute.

In order to get the political support, we need to make this economic shift, broad and cohesive peoples' movements are needed. We must unite in our call to reverse corporate deregulation. This can be accomplished through what I call "big-picture activism." Big-picture activism highlights the interconnections between the economic, ecological, and humanitarian issues we care about. It focuses on our shared roots in an out-of-control global economic system. It means actively spreading information about how scaling down, slowing down, and localizing economic activity is an effective and strategic "solution multiplier."

While COVID-19 has awakened many people to the problems of globalization, I do not believe that it will be enough to halt this global juggernaut. In order to sustain a truly transformative movement, we need to generate greater, more widespread, and more holistic awareness about the flaws of the global economy. We can once again embrace the living, breathing, systemic alternative that is economic localization. We need to link hands across the ecological and the social, the left and the right political divides, in order to restore the economic structures necessary to meet our needs and nurture the only planet we have. The good news is that the potential to build a united movement—one powerful enough to bring about fundamental change—is greater than ever before. We have an opportunity now to come together to bring about total transformation.

References

Boers, Elroy, Mohammad H. Afzali, Nicola Newton and Patricia Conrod. "Association of Screen Time and Depression in Adolescence." *JAMA Pediatrics* 173.9 (2019) 853–59.

Gardner, Matthew. "Amazon in Its Prime: Doubles Profits, Pays $0 in Federal Income Taxes." *Just Taxes Blog*, February 19, 2019. https://itep.org/amazon-in-its-prime-doubles-profits-pays-0-in-federal-income-taxes/.

Local Futures. "Let's Localize Like Never Before: Answering the Call of COVID-19." Accessed August 11, 2020. https://www.localfutures.org/covid-19/#1585861956304-f44ed519–4eac.

Opinion Has It. "Our Digital, No-Touch Future." Project Syndicate, June 2, 2020. Podcast, 27:26. https://www.project-syndicate.org/podcasts/our-digital-no-touch-future.

Pariser, Eli. "Beware Online 'Filter Bubbles.'" Filmed March 2011 in Long Beach, CA. TED video, 8:49. https://www.ted.com/talks/eli_pariser_beware_online_filter_bubbles#t-5116.

25

Whakawhanaungatanga as a Blueprint for Radical Societal Transformation

Rebecca Kiddle

We hear a lot of talk about the word *community*. There seems to be general acceptance that a strong community is good for us. But seeing it realized, and being part of something like that, seems more elusive for many of us than it has ever been. The current tumult caused by COVID-19—along with the anti-racism movement in the US and other places like Aotearoa, New Zealand from where I write—have encouraged many of us to reflect on key ideas that shape our world. I don't think I am the only one who is thinking about community. This time of reflection has led us to question whether we are truly a part of community. Is community something that is important for our lives? Which communities are able to live privileged and peaceful existences, and which aren't? Do our communities really allow for difference? Or are they actually microcosms of race and class that lead us only to interact with those who look and act, and are in fact just like us.

Lockdown in New Zealand

During lockdown in Aotearoa, New Zealand, we were asked to stay within our geographical communities. As a result, people took to the streets, local parks, and green spaces more than ever before. We craved "connection with others" beyond our four walls. We started inhabiting the streets and roads of our neighborhoods in ways that most of us had never seen before. Our city center was eerily quiet. Across the globe, images of animals re-inhabiting places now devoid of cars plastered our TV screens. Many of us marveled at just how quickly these non-humans re-colonized spaces once taken by concrete mixers, bricklayers, and other instruments of human domination.

Suburban neighborhoods seemingly thrived with people holding church services and exercise classes in the street while maintaining social distancing. In New Zealand, we congratulated ourselves as the "team of five million" who worked together to achieve (at one point) a COVID-free country. This idea of team evoked a strong sense that we are all valued members of community—even if a socially constructed, imagined community.

As an introvert, I relished the opportunity to stay home. I am able to conduct all of my well-paid employment as a university academic online— even if it did create a little stress having to learn new digital tools along the way. My days were spent recording lectures and coordinating online architectural tutorials and assessments, punctuated with a dog walk each day. Early on in lockdown I picked up an order of freshly baked croissants from a nearby fancy bakery, along with ten kilograms of bakers' flour. Like many others, I also took the opportunity to learn new baking skills and try my hand at new bread recipes. I gave my best effort to use this flour well.

My privileged existence in a warm, dry home with fast Wi-Fi, no children climbing over my shoulder to disrupt a work Zoom meeting, and the certainty of ongoing income was not lost on me. News stories came across our TV screens of a sharp increase in requests for food bank provisions. The ongoing loss of employment for many of those on casual contracts, or who had jobs that were made precarious by COVID-19, left a lasting mark. Meanwhile, many essential workers have had to put themselves in perilous situations, despite their minimum wage contracts, to ensure that the rest of us would be fed. Even again this morning when I turn on the TV, the news is highlighting countries that are experiencing deep poverty, showing how COVID-19 is further exacerbating the harsh realities of their existence.

Herein lies the conundrum. My meager attempts at connection through a two-meter distanced "hello" during my dog walks were often met with a tense, head down, fear-filled response. My efforts to reach out were matched with concern that my two-meter-distant hello might still somehow transfer the dreaded illness. The WhatsApp group created by a neighbor for those on my street, far from creating deeper community, acted as little more than a means for identifying misdirected mail.

Māori Self Determination

At the national level, reports started to surface that showed how Māori, the Indigenous people of New Zealand, were policed at far higher rates during lockdown than white New Zealanders (*Pākehā*).[1] Our team of five million, our imagined community, seems to have some cracks in it.

Still, in some parts of New Zealand community was being exemplified in hopeful and self-determining ways. Māori tribal groups on the east coast of Aotearoa and further north in Northland communities were mobilized by local tribal leaders to help stop the spread of the virus. Distant memories of the Spanish flu that had wiped out large numbers of Indigenous people in Aotearoa—at rates some eight times higher than that of the *Pākehā* or white New Zealanders—crashed through into our current reality. Echoing these memories, we learned how far contemporary health outcomes for Māori lag behind the rest of the New Zealand population. This information heightened tribal leaders' desire to ensure that every effort was being made to protect Māori.

Community checkpoints were enacted by Māori as a form of protection. Tribal communities set up rosters for people to service the key entry points into their regions. They talked with anyone traveling by car, communicating the risks to the community of new arrivals. Traveling to these remote communities happened when people were not supposed to leave their neighborhoods, so the trips were occurring against the policy of the day.

The act of setting up these community checkpoints asserted the right of self-determination for Indigenous communities. The right to take control of their own destinies is called *mana Motuhake* in the Māori language. It's a right afforded Māori under the Treaty of Waitangi, a treaty signed between the British Crown and Māori chiefs in 1840 when the British formalized the colonization of New Zealand. However, this right has since been systematically undermined; and many, including the New Zealand government, disregard the right of Māori to make decisions for their own families and communities

Many have argued for and against the legality of these checkpoints. The opponents suggest the Māori were unlawful and or that they were "acting above their station."[2] In response, Māori have asserted the need to protect particularly vulnerable communities. Incidentally, widespread gratitude

1. Daadler, "Māori Over-represented."
2. Harawira, "Who the Hell Do You Māoris."

has been shown for these checkpoints by non-Māori in the community.[3] My goal in this chapter is not to debate who was right—though two legal scholars, Max Harris and Professor Emeritus David V. Williams, have offered compelling arguments for the legality of these checkpoints.[4] Instead, I would like to explore the underlying principle that brought this group of people together to strengthen and protect the wider community, not just Māori—the principle of *whakawhanaungatanga*.

Māori Community

In te reo Māori (the Māori language), *whakawhanaungatanga* is defined as:

> relationship, kinship, sense of family connection—a relationship through shared experiences and working together which provides people with a sense of belonging. It develops as a result of kinship rights and obligations, which also serve to strengthen each member of the kin group. It also extends to others to whom one develops a close familial friendship or reciprocal relationship.

When you put the prefix *whaka* in front of a word in te reo Māori, it generally causes something to happen. *Whakawhanaungatanga* refers then to the *process* of establishing and building relationships. Relatedly, the Māori dictionary suggests that this process should happen in a qualitatively good fashion, that is, it involves "relating well to others."[5] In the Māori worldview, the process of getting to an outcome is of utmost importance. This is seen most clearly in our welcome rituals, *pōwhiri* and *whakatau*. In these rituals, newcomers to a tribal area must undergo a process of moving from being *waewae tapu* (sacred feet) to *noa* (profane). The same can be said for community. Establishing and then maintaining relationships qualitatively well sits at the heart of all interactions in the Māori worldview.

Community can happen simply through a co-location of people in a geographical neighborhood. These communities self-select as they tend to attract people of similar economic, ethnic, and racial groupings. This can make for tight-knit monocultural communities but doesn't typically provide the base for strong multicultural communities.

---

3. Scoop Independent News, "Widespread Gratitude."
4. Harris and Williams, "Community Checkpoints."
5. Māori Dictionary, "Whakawhanaungatanga."

One of my recent thesis students, Emma McNeill, has done some work to help with my project "Where Do We Dance."[6] In this project, we look at the spaces where New Zealanders come together to build community. Emma found that despite Newtown (a neighborhood in Wellington, the capital city of New Zealand) being one of the most diverse suburbs in town, those outside of the dominant cultural group required some kind of "in" to acculturate with locals.[7] In addition, when contact did happen it tended to be a one-way street, with those from the dominant white culture not feeling compelled to engage with other cultural groups.

The "in" tended to be an event or activity such as a shared lunch, but generally providing physical space for such interaction was not enough. People did not naturally interact across cultural groups despite perceptions of Newtown as a hive of multiculturalism. It gradually became clear to us that Newtown is apparently a neighborhood that houses a range of distinct mini-communities under a veneer of multiculturalism.

I suggest that the *whaka* part of *whakawhanaungatanga* is crucial if we are ever to move beyond needing to continuously assert that "black lives matter." Our societies need to begin offering opportunities to be part of communities that—at the very least—can cope with diversity, while ideally offering people opportunities to build strong ties to one another.

Whakawhanaungatanga then is key. In the case of the recent COVID-19 community checkpoints, Māori knew from history that they tended to fare worse off than non-Māori in pandemic situations. So while Māori set about providing checkpoints primarily for the sake of their elderly, I would argue they were also doing it for the broader community as an act of *whakawhanaungatanga*. These groups just did it. Government agencies, such as the police and local government, fell (somewhat) in line to support them, when they could see that these groups were on to a good thing.[8] The checkpoints worked to hold off COVID-19 from these particularly vulnerable communities.

Living for the Benefit of All

Imagine what could be possible if the government post-COVID (if there is ever going to be a time such as this!) honored the Treaty of Waitangi and gave Indigenous peoples like us the agency and resources to self-determine our own destinies for our own good and the good of others.

6. Where Do We Dance, "Survey."
7. McNeill, "Spaces of Cross Cultural Encounters."
8. Smith, "Community 'Checkpoints.'"

At all levels of government, legislation and policy needs to push for these initiatives. They need to give power and resources to Mana Whenua (the particular tribe of that region that has guardianship over that land) to enact *whakawhanaungatanga* for the benefit of all of us. Many tribal groups have lamented the loss of resources. This is seen most painfully in the destruction of food gathering places through pollution, and in the wanton urban development that is led and allowed by government. Our lament isn't just for these resources; it's also because this loss has taken away the ability to show *manaakitanga* (hospitality) to others.

Decisions to hold on to power are rooted in a system that doesn't trust Indigenous people to carry out the clear *tikanga* (customs) that are important to us. This seems highly ironic to me, given that *Pākehā* New Zealanders themselves have no obvious or codified rules for showing hospitality to "outsiders." In fact, it could be said that the problem isn't just the lack of rules or processes to include others, but rather the historical and contemporary conditioning that values assimilation to *Pākehā* ways of being. Their way of being in the world is dependent on the erasure of others, to the point that we must now assert "black lives matter."

Whakawhanaungatanga allows for difference. It gives everyone the opportunity to say who they are, and for those listening to make a connection to whoever has the floor. It provides a formal "in" that is needed to give all persons the space to talk. This can take time. I've been in *whakawhanaungatanga* sessions which took hours, only stopping in the wee hours of the morning. In these sessions everyone had the opportunity to contribute their identities to the rich mix in the room.

What might a government policy look like that is based on *whakawhanaungatanga*? Or even better, what might our societies and communities look like if Māori were reinstated with the resources that were unlawfully taken away in 1840? And what if Māori were given agency to enact our own processes of *manaakitanga* (hospitality), through *whakawhanaungatanga*, so that we could again build strong communities that are happy with difference? To my mind, the community checkpoints offer a microcosm of the problem, and give an inkling of what is possible.

Now that the initial burst of enthusiasm for "our team of five million" has faded, we can see that we have much work to do. We can see that even our most diverse suburbs aren't the multicultural hubs we once thought. If we want to move forward with a real sense of community, we must learn to practice *whakawhanaungatanga* together.

References

Daadler, Marc. "Māori Over-represented in Lockdown Police Proceedings." *Newsroom*, June 3, 2020. https://www.newsroom.co.nz/2020/06/03/1216951/maori-overrepresented-in-lockdown-police-proceedings?fbclid=IwAR2G9REdDQIWuA zsIN8vfdNeKi5zD1kI1lUd0-r1N6cPO3mYbjuTQUamplE.

Harawira, Hone. "Who the Hell Do You Māoris Think You Are?" *Northland Age*, May 14, 2020. https://www.nzherald.co.nz/northlandage/news/article.cfm?c_ id=1503402&objectid=12331881.

Harris, Max, and David V. Williams. "Community Checkpoints are an Important and Lawful Part of NZ's Covid Response." *Spinoff*, May 10, 2020. https://thespinoff. co.nz/society/10-05-2020/community-checkpoints-an-important-and-lawful-part-of-nzs-covid-response/.

Māori Dictionary. "Whakawhanaungatanga." Accessed August 10, 2020. https:// maoridictionary.co.nz/search?idiom=&phrase=&proverb=&loan=&histLoanWord s=&keywords=whakawhanaungatanga.

McNeill, Emily. "Spaces of Cross Cultural Encounters." Masters' thesis, Victoria University of Wellington, 2019.

Scoop Independent News. "Widespread Gratitude for Roadside Checkpoints." Accessed August 10, 2020. https://www.scoop.co.nz/stories/AK2005/S00047/widespread-gratitude-for-roadside-checkpoints.htm.

Smith, Mackenzie. "Community 'Checkpoints' Credited with Reducing Covid-19 Spread." *Radio New Zealand*, April 6, 2020. https://www.rnz.co.nz/news/national/413545/ community-checkpoints-credited-with-reducing-covid-19-spread/.

Where Do We Dance. "Survey." 2020. https://wheredowedance.com.

26

Now That We Know the Critique of Global Capitalism Was Correct

Arturo Escobar

THE COVID-19 CRISIS HAS brought to the fore a renewed awareness of the possibility—for many, the absolute imperative—of a radical eco-social, economic, political, and cultural transition in every country and in the world at large. This sense has found powerful expression in a slogan that has been circulating in Latin America in relation to the crisis: "*No volveremos a la normalidad, porque la normalidad era el problema*" (We shall not go back to normalcy, because normalcy was the problem to begin with). A recent "Eco-Social Pact from the South," issued by Latin American social movements and intellectuals, attempts to give form to this principle.[1]

In this chapter, I would like to share my own sense of strategies for such transition, in the context of the pandemic and beyond. Transitions and transition design have been an active focus of research and practice for me since I wrote *Designs for the Pluriverse*.[2] Such transitions have been taking place throughout Latin America, particularly through projects in Colombia and, more recently, in the context of the intense debates on how to design the post-pandemic world. While many narratives of transition have focused on the relation between the pandemic and capitalism, the most interesting

1. Pacto Ecosocial Del Sur, "Pacto Ecosocial."
2. Escobar, *Designs for the Pluriverse*.

247

ones, in my view, tackle this relation through the broad lens of civilizational ruptures and transitions, which have also been prominent in Latin America.

What follows is a brief summary of the argument that I have advanced. It takes the form of suggesting five "axes"—principles for thinking about strategies for the transition, whether through design or through other forms of collective action. Each of these five axes connects with the pressing issues and open questions in transition studies and design, including ontological-ly-oriented inquiries.

(1) Returning the Communal to Social Life

This first action-focused orientation begins with a resounding *no* regarding individual solutions to the current pandemic or as-yet-unknown future crises. We must actively and fully resist modern capitalism's ever more efficient way of making us feel as if we are individuals isolated from family, kin, and society. Intent on creating consumers who see themselves as individuals making decisions solely in market terms, globalization has entailed an un-compromising war against everything that is communal and collective. Yet history teaches us that human experience has largely been placed-based and communal, lived at the local level.

A locally-oriented life is one lived in relationship with the humans and other species around us. This kind of life responds to (and helps foster) a symbiotic co-emergence of living beings and their worlds, which results in what Gutierrez Aguilar calls "communitarian entanglements" that make us kin to all that is alive.[3]

Oaxacan activists refer to this dynamic as the "*condición nosótrica de ser*" (the we-condition of being). If we see ourselves *nosótricamente,* we will of course adopt the principles of love, care, and compassion as our ethical framework for living, starting with our home, place, and community. We do this not to isolate ourselves but to prepare for a *greater* sharing of our lives, one rooted in autonomy and communication. We bring this orientation of *compartencia* ("sharingness") to our thoughts and actions when our goal is to deepen our understanding of what makes for a resilient community (or a resilient person), or when we imagine creating whole new communities.

---

3. Gutiérrez Aguilar, *Horizontes Comunitario-populares.*

(2) Returning the Local to Social,
Economic, and Cultural Activities

Historically, human communities have constantly moved and regrouped. With global capitalism and development, however, the pressures that undercut local communities, and the unchosen movement of people away from their local lands—often imposed by force, as with slavery—has increased exponentially. We see this in the dramatic ways that people and their communities are forced from their land due to large-scale projects to extract resources, such as logging in rain forests or mining in Indigenous lands. Given the high social and ecological costs, we need to oppose the pressures that delocalize people in order to extract more natural resources.

The COVID pandemic is fostering a new awareness that capitalist globalization is not inevitable, even when our survival as individuals and as a species seems threatened. As Gustavo Esteva has argued, COVID is reestablishing the importance of the local.[4] Regaining our rootedness in the local means relocating essential life activities back into the places where we live. Food is one such crucial area; it's also where communitarian and locally-focused innovation is already occurring. Food sovereignty, agroecology, seed saving, commoning, and urban gardens are instances of this new turn "back to the local." Note that all of these are also innovations that break with patriarchal, racist, and capitalist ways of living. Applied at the local level, these and similar changes can foster transformations of national and international food production systems. For example, they could lead to a renewed understanding of the value of commonly held land, or to re-weaving the ties that once flourished between cities and the surrounding countryside.

"Returning to the local" means thinking through a range of strategies that are formulated as active verbs—to eat, to learn, to heal, to dwell, to build, to know—as opposed to passive services provided *for* us, such as health, education, food, housing. This new way of thinking can be the impetus for a significant re-orientation of how we and our communities think about and understand the many worlds we inhabit, including the domains of the physical, family, society, and many others.

(3) The Strengthening of Autonomies

Autonomy refers to the political aspect of returning to the communal and the local. Without autonomy, movement toward the local will only go

---

4. Esteva, "El Día Después."

half-way, or it will be too easily reabsorbed by newer forms of delocalized re-globalization.

There has been a vibrant debate on *autonomía* in Latin America since the 1994 Zapatista uprising. At times, autonomy is thought as the radicalization of direct democracy. In fact, though, it offers a new way of conceiving and enacting politics. Politics, on this view, has three dimensions: it is the inescapable task that emerges from the entanglement of humans among themselves and with the earth; it is oriented to reconfiguring power in less hierarchical ways; and it uses principles such as sufficiency, mutual aid, and the community's self-determination of its own norms of living.[5] In many parts of the world, autonomy is at the heart of a great deal of political mobilization; it also undergirds many less openly political practices. At its best, autonomy is *the theory and practice of inter-existence*. It's about designing for and with what we call the "pluriverse," which means a world in which many worlds fit together harmoniously.

These first three strategies aim to create dignified lives for people in each region. They require us to rethink or reimagine the economy in terms of everyday solidarity, reciprocity, and conviviality. There are many clues for this project among those groups who, even during the COVID-19 pandemic, have remained dedicated to living well—constructing instead of destroying, reuniting instead of separating. The strategies offer tangible and actionable principles for "dream-designing" (*disoñación*), helping us to redesign our lives in a partial, but still substantial movement toward deglobalization. We dream of the end of globalization as we know it, or at least of the beginning of a globalization in different terms—a globalization based on the paradigm of care (*cuidado*).[6] (Of course, in this case we may no longer want to call it "globalization.") We dream of a world in which there is place for many worlds—a pluriverse.

When we return to the local and the communal, we must also return the economy to its proper place in our lives. In the modern era, capitalism made the economy central to our lives, but it also separated the economy from the families, communities, and places where we make our home. Can we redesign our local, regional, national, and even transnational economies to favor relationships, the commons, and our lives? Can we take the economy with its capitalist orientation out of our social lives? Can we even remove profit motives from our labor and markets?

5. Understanding the threats to autonomy requires us to think about strategies of "overturning and flight" in relation to the established orders of capitalist modernity and the state. See Gutiérrez Aguilar and Skar, *Rhythms of the Pachakuti*, 41.

6. Svampa, "Reflexiones."

(4) Transforming the Impact of Race, Patriarchy, and Colonialism

Patriarchy is so entrenched in our personal thoughts and desires, in the values and expectations of our social lives, and even in our very civilization, that it may seem impossible to dismantle it, much less transform it. If we are to inhabit new ways of living, however, we must first identify, question, and challenge the patriarchal assumptions that seem so natural to us and that are such a deep part of our daily lives.

Latin American feminists remind us of how high the stakes are. As many have written, there is no way to end colonization without ending the patriarchy and the racializing of social relations. To end these, however, means learning how to practice a "politics in the feminine."[7] This practice centers on the reproduction of life (in the broadest sense) and on restoring the focus on collectively produced goods. Think, for example, of the work of the Afro-Colombian philosopher Elba Palacios, who work with poor Afro-descendant Black women in Cali, Colombia. Palacios follows the peaceful routes of Colombian women with racial and ethnic identities, observing how they reconstitute their territories and maintain dignified lives. "In their territories," she writes, "women give birth to life and to modes of re-existence." According to this activist-researcher, the women teach us that "to re-exist means much more than resist. It involves the creation and transformation of autonomy in defense of life, through a sort of contemporary urban maroonage that enables them to reconstitute their negated humanity, reweaving communities in the historical diaspora."[8]

I also believe that this feminist and antiracist lens is essential if we are to understand (and strengthen) the process of relocalizing and reestablishing thriving communities in many different parts of the world.[9]

How do we overcome patriarchy and the racialization of social existence? We must learn to repair and to heal the tapestry of interrelations that makes up the bodies, places, and communities that we are and that we inhabit. Here we can learn from the diverse movement of communitarian feminisms led by Mayan and Aymara activist-intellectuals.[10] Gladys Tzul Tzul, for example, highlights the potential of communal life as the horizon

7. More technically, to depatriarchalize and deracialize requires repairing the damage caused by the heteropatriachal white capitalist ontology. See Segato, *La Guerra Contra*; Segato, *Contra-pedagogías*; and Gutiérrez Aguilar, *Horizontes*.

8. Palacios Cordoba, "Sentipensar la Paz."

9. See also Lozano Lerma, *Aportes a Un Feminismo*; and Hartman, *Wayward Lives* for the historical experience of young Black women in the US.

10. See also Tzul Tzul, *Sistemas* and Julieta Paredes, *Hilando Fino.*

for understanding the struggle and as a natural space for the continuous reconstitution of life. Her perspective is deeply historical and anti-essentialist as she explores the complexity of thinking from and living within communitarian entanglements (*entramados comunitarios*), together with all the forms of power that run through them.[11] From this perspective, we can see that reconstituting life's web of relations in a communitarian manner is one of the most fundamental challenges faced by any transition strategy. The Argentinean anthropologist Rita Segato has put this point powerfully:

> We need to advance this politics day by day, outside the State: to re-weave the communal fabric in order to restore the political character of domesticity, which is a core feature of communal life . . . *To choose the relational path is to opt for the historical project of being community* . . . It means to endow relationality and the communal forms of happiness with a grammar of value and resistance—a grammar that is able to counteract the powerful developmentalist, exploitative, and productivist "rhetoric of things" with its alleged meritocracy. This strategy, from now on, is a feminine one [*la estrategia a partir de ahora es femenina*].[12]

This feminist and radically relational politics need to be incorporated into many, if not all, of our designing practices; indeed, it needs to be designed into our anthropology itself.

(5) The "Re-earthing" of Life

We arrive, finally and necessarily, at the Earth (Gaia, Pachamama, co-emergence, self-organization, symbiosis). I speak of Earth in the broadest sense, drawing on Indigenous cosmological visions as well as on contemporary scientific findings and theories. There exists a core, indubitable fact: everything exists because everything else does. We live on a planet that exhibits a profound interdependence. Nothing exists apart from the geologic eras and biological evolution that proceeded it.

Life is an unceasing unfolding of changing forms, behaviors, and relations. New forms of life are always in the process of co-arising. The great

11. Contrary to common thinking, Indigenous communitarian formations are not homogeneous but plural; neither do they suppress personal expression: "The communal does no place limits on the personal, it rather potentiates it. The communitarian entanglements provide the grounds on which personal and intimate lives are sustained" (Tzul Tzul, *Sistemas*, 57), even if the organization of life, of politics and of the economy is realized collectively and every family has to engage in these practices.

12. Segato, *La Guerra Contra Las Mujeres*, 106 (my emphasis).

biologist Lynn Margulis goes beyond this simple fact when she writes, "Gaia, as the interweaving network of all life, is alive, aware, and conscious to various degrees in all its cells, bodies, and societies."[13] I hold this notion of an ever-changing Earth whenever I try to put theory into practice. Indeed, our evolving notions of sustainability are one crucial way that we express our desire to return to a more Earth-focused existence.

Another way is lived out by the Nasa Indigenous people of Northern Cauca in Colombia's southwestern region. For nearly two decades, their social and communitarian Minga has articulated the concept and the movement of the liberation of Mother Earth, as part of their strategy of "weaving life in liberty." As they say, Earth has been enslaved, and as long as she is enslaved, all living beings on the planet are also enslaved. The Nasa's struggle involves both the active recovery of their traditional lands *and* a different mode of existence on the reclaimed lands. "This struggle," they say:

> comes out of Northern Cauca, but it is not Northern Cauca's struggle. It comes from the Nasa people, but it is not the Nasa people's. Because life itself is at risk when the earth is exploited in the capitalist way, which throws the climate, the ecosystems, everything out of balance.

As they hasten to clarify, it's a project for everybody, since we are all Earth and pluriverse: "Every liberated farm, here or in any corner of the world, is a territory that adds to reestablishing the balance of Mother Earth (*Uma Kiwe*). This is our common house, our only one. Yes, indeed: come on in, *the door is open*."

What does it mean to accept this invitation—whether in the countryside or in the city, in the Global South or the Global North? The liberation of Mother Earth, conceived from the cosmocentrism of peoples such as the Nasa, invites us to dream and to imagine (*disoñar*: "dreamagine") different ways of designing and building worlds. What we can learn from them are ways of dreaming-and-imagining how to reconstitute the entire web of life, how to sustain the regions and territories in which each of us lives, and how to develop communal forms of economy—wherever we are.[14]

---

13. Margulis, *Symbiotic Planet*, 126.

14. There is an extensive Nasa archive on the liberation of Mother Earth. See, for example, ACIN, "Libertad para la Madre Tierra"; ACIN, "El desafío que nos convoca"; Tejido de Educación ACIN, "Lo que vamos aprendiendo"; and Almendra, "La paz de la Mama Kiwe." I must emphasize that the movement for the liberation of Mother Earth is currently divided and, at the same time, heavily repressed by landowners and government forces. For a full account of the situation, see Escobar, *Pluriverse Politics*, ch. 3.

The liberation of Mother Earth is what we call a "social imaginary" for peoples and collectives, wherever they happen to be. It's also not as utopian a project as it might seem. For historical reasons, Latin America has been preparing for this fundamental project at many levels. Little by little, Latin Americans have been generating an entire space of knowledge, being, and politics where Earth-centered struggles and forms of knowing all converge. This convergence has become more noticeable in the wake of the many struggles that have been triggered by the people's resistance to the brutal extraction of natural resources over the past few decades. As a space of resistance, it is grounded in the conviction that the devastation of the planet is not an inevitable destiny.

More recently, the Earth question has attained an incredible urgency. It is powerfully expressed by environmental scholars such as Mexican Enrique Leff and Colombian Ana Patricia Noguera. Noguera, for example, studies the "geometries of atrocity produced by a calculating world." She shows how Latin American and Abya-Yalan environmental thought responds with a "geo-poetics of weaving-dwelling" that is oriented towards the original modes of inhabiting the planet.[15] These philosopher-activists succeed at reframing environmental discourse as Indigenous, Black, peasant, feminist and ecological struggles. What they offer as a result is an entire archive of categories and practices that can help us find fruitful paths to the concrete transitions that humanity must now make.

Concluding Thoughts

"When we think with the audacity of world builders," the practitioner-theorists from the Design Studio for Social Intervention in Boston tell us, "we begin to see not just new ways of fighting for a more just and vibrant society, but whole new ideas about what that world might be like."

So let's imagine ourselves as designers of everyday life—as people who can begin to imagine what is possible, and then bring it into the reach of practice. Let's ask: What profound rearrangements are we yearning for? What *could* our families, our communities, and our daily lives be like? Let's heed this question with all our hearts and minds. Let's join our deepest yearnings with those of others, and together create a new world with new ways of being.

15. Noguera et al., "Metodoestesis."

References

Almendra, Vilma. "La paz de la Mama Kiwe en libertad, de la mujer sin amarras ni silencios." Pueblos en Camino. August 2, 2012. http://pueblosencamino.org/?p=150.

Asociación de Cabildos Indígenas del Norte del Cauca (ACIN). "El desafío que nos convoca." May 28, 2010. http://www.nasaacin.org/el-desafio-no-da-espera.

———. "Libertad para la Madre Tierra." 28 Mayo 2010. http://www.nasaacin.org/libertar-para-la-madre-tierra/50-libertad-para-la-madre-tierra.

Escobar, Arturo. *Designs for the Pluriverse: Radical Interdependence, Autonomy, and the Making of Worlds*. New Ecologies for the Twenty-first Century. Durham: Duke University Press, 2018.

———. *Pluriversal Politics*. Latin America in Translation. Durham: Duke University Press, 2020.

Estiva, Gustavo. "El Día Después." *La Jornada*, April 6, 2020. https://www.jornada.com.mx/2020/04/06/opinion/020a2pol.

Gutiérrez Aguilar, Raquel. *Horizontes Comunitario-populares: Producción De Lo Común Más Allá De Las Políticas Estado-céntricas*. Madrid: Traficantes De Sueños, 2017.

Gutiérrez Aguilar, Raquel and Stacey Alba D. Skar. *Rhythms of the Pachakuti: Indigenous Uprising and State Power in Bolivia*. New Ecologies for the Twenty-First Century. Durham: Duke University Press, 2014.

Hartman, Saidiya V. *Wayward Lives, Beautiful Experiments: Intimate Histories of Social Upheaval*. First ed. New York: Norton, 2019.

Lozano Lerma, Betty Ruth. *Aportes a Un Feminismo Negro Decolonial: Insurgencias Epistémicas De Mujeresnegras-afrocolombianas Tejidas Con Retazos De Memorias*. Serie Investigación Decolonial 1. Quito: Abya Yala, 2019.

Margulis, Lynn. *Symbiotic Planet: A New Look at Evolution*. Science Masters Series. New York: Basic Books, 1998.

Noguera, Ana Patricia, Leonardo Ramírez, and Sergio Manuel Echeverri. "Metodoestesis: Los Caminos Del Sentir en los Saberes de la Tierra un Aventura Geo-Epistemica en Clave Sur." *Revista De Investigación Agraria Y Ambiental* 11.3 (2020) 45–62.

Pacto Ecosocial Del Sur. "Pacto Ecosocial Del Sur." Accessed August 10, 2020. https://pactoecosocialdelsur.com/#1592362596334-8e141cec-613c.

Palacios Cordoba, Elba Mercedes. "Sentipensar la Paz en Colombia: Oyendo las Reexistentes Voces Pacíficas de Mujeres Negras Afrodescendientes." *Memorias* 38 (2019) 131–61. http://dx.doi.org/10.14482/memor.38.303.66.

Paredes, Julieta. *Hilando Fino: Desde El Feminismo Comunitario*. La Paz: CEDEC: Comunidad Mujeres Creando Comunidad, 2008.

Segato, Rita Laura. *Contra-pedagogías De La Crueldad*. Argentina: Prometeo Libros, 2018.

———. *La Guerra Contra Las Mujeres*. Madrid: Traficantes De Sueños, 2016.

Svampa, Maristella. "Reflexiones Para un Mundo Post-Coronavirus." *Nueva Sociedad*, April 2020. https://nuso.org/articulo/reflexiones-para-un-mundo-post-coronavirus.

Tejido de Educación ACIN. "Lo que vamos aprendiendo con la liberación de Uma Kiwe." Pueblos en Camino. January 19, 2016. http://pueblosencamino.org/?p=2176.

Tzul Tzul, Gladys. *Sistemas De Gobierno Comunal Indígena: Mujeres Y Tramas De Parentesco En Chuimeqèna'*. Guatemala: SOCEE, Sociedad Comunitaria De Estudios Estratégicos : Tz'i'kin, Centro De Investigación Y Pluralismo Jurídico: Maya' Wuj Editorial, 2016.

TOMORROW

Rise
by Nina Montenegro

27

Telling a New Story

David C. Korten

A Species in Terminal Crisis

On October 1, 2018, the UN Intergovernmental Panel on Climate Change issued a warning to the world. To prevent irreparable consequences for Earth and humanity, we need to reduce greenhouse gas pollution by forty-five percent from our 2010 levels by the year 2030, and reduce these gases by one hundred percent by 2050. Just a little over a year after that warning, the deadly COVID-19 pandemic forced a global economic shutdown that revealed gaping institutional failures and injustices. In the midst of the pandemic, the video of the intentional murder of an unarmed young black man by a white police officer triggered mass protests against systemic racism across the US and the world. We can no longer deny the massive failure of established institutions.

According to estimates of the Global Footprint Network, humans—in exceedingly disproportionate respects—consume at a rate 1.75 times what Earth's regenerative systems can sustain.[1] Our excess consumption is disrupting the living systems that provide the food we eat, the water we drink, and the air we breathe. These systems restrain rogue viruses and stabilize the climate that shapes the daily life of every living being.

1. Global Footprint Network, "Earth Overshoot."

As the collapse of our interlinked environmental and social systems render ever more places uninhabitable, millions abandon their homes in fear and sadness to seek refuge in what remains of Earth's livable plots of land. Simultaneously, growth in extreme inequality reduces the vast majority of the world's people to a daily struggle for survival. According to Oxfam, the combined financial assets of twenty-six billionaires are now greater than those of the poorest half of humanity.[2] In the United States, the collective assets of wealthiest one percent exceed those of the bottom ninety percent.

The failures of our global economy have been accumulating year after year, decade after decade. Despite the failures many of us see, feel, and live every day, public support for the current economic system is maintained through a set of ideas that promises a better future. The values are variously formulated: "Money is the defining human value. Growing GDP is the defining goal of government and growing profits the defining goal of business. Pursue these goals and all people will ultimately enjoy a future of comfort and material affluence."

This is a set of ideas crafted and legitimated by academic economists with support from the same wealthy patrons whose interests they serve. It doesn't stop with university professors. These wealthy patrons support the continual retelling of this story through institutions of education, media, politics, and religion. Many of us focus so ardently on the relentlessly promoted promise that we fail to notice the deep conflict contained within the story that drives the global economy. These incompatibilities can be summarized by noting three foundational truths:

1. *Human wellbeing depends on the wellbeing of the living Earth.* All living beings depend on living communities that self-organize to create, share, and continuously regenerate through their labor the conditions essential to life. We are all children of the living Earth that birthed and nurtures us. Our wellbeing depends on her wellbeing. She long existed without us, but we cannot exist without her. Restoring her health through the wise application of our labor must be a defining economic priority.

2. *Humans are a choice-making species of many possibilities.* The immense diversity across the span of history of human cultures and institutions demonstrates that we are a species of many possibilities. We can, for example, cooperate and nurture. Or compete and exploit. What defines our distinctive nature is our ability to make shared cultural and

2. Elliot, "World's 26 Richest."

institutional choices that in turn shape our individual and collective relationships with each other and Earth.

3. *The drive to grow money imperils the human future.* Money is a number that has value only when other people have something to sell that we need or desire. Useful as a tool, money becomes dangerous when embraced as a purpose. A society that chooses to exploit people and nature for the sake of money destroys Earth's capacity to support life and leads ultimately to human self-extinction. These unbearable costs are usually undertaken so people who already have more than they need can continue to accrue, and the gap between the rich and poor expands.

A Shared Story

Humans are distinctive among Earth's species. We organize around shared cultural stories of our origin, nature, and purpose. These stories become the lens through which we see our world. They help us define the values and institutions that mold our relationships with one another and the Earth. The societies we have imagined, dreamed, and built range from being characterized by loving cooperation to ones characterized by violent competition. These diverse results reveal the power of story. They explore the extraordinary range of human possibility and plum our potential to choose our future.

Get our story right, and we flourish together in the service of life. Get it wrong, and we become an existential threat to ourselves and to the Earth that graciously birthed and nurtures us. At present we live in the grip of a deeply flawed story. To change the situation, we must find our way to an authentic narrative. We need a story informed by traditional wisdom, the world's great religious traditions, and the leading edge of science.

Humans have long dreamed of a thriving world filled with communities that offer ecologically balanced and spiritually fulfilling lives. Africans move within a spiritual heritage of *ubuntu*, often translated as "I am because you are." The Quechua peoples of the Andes talk about it as *sumac asway*. It translates into Spanish as *vivir bien* and into English as good living. Bolivia and Ecuador have etched this concept into their constitutions. China has written it into its constitution as a commitment to an ecological civilization. In 2015, the Parliament of the World's Religions issued a Declaration on Climate Change that closed with these words: "The future we embrace will be a new ecological civilization and a world of peace, justice, and sustainability, with the flourishing of the diversity of life. We will build this future as one human family within the greater Earth community."

Far from being a call to sacrifice, these challenging times call us to actualize the potential of our human nature and our deep inclination to love and to care for one another and the Earth. Together we can embrace the requirement to significantly reduce total human consumption. We can choose our current moment as an opportunity to relieve ourselves and the Earth from the enormous environmental and social burdens imposed by war, obsessive materialism, planned obsolescence, and infrastructures of self-reliance that separate us from one another and nature.

Even if GDP and corporate profits decline, this need not be our primary concern as long as we correct the institutional flaws. In a system designed to crash if money does not continuously flow from the poor to the rich, we must see our situation for what it is. We can refuse this hostage situation in which no one ultimately wins, where all remain captive to GDP. We can say no to these mechanisms of manipulation.

Awakening from a Deadly Deception

Humanity's existential crisis traces—at least in part—to mainstream economics: a political ideology posing as an objective science. With generous financial support from the powerful financial interests, its favored story has largely defined humanity's purpose and dominant institutions since the mid-twentieth century.

Because this ideology has been presented in most of the world's universities as uncontested truth, generations of leaders have been taught to believe that financial assets are the measure of a society's worth. Therefore, supporting growth of these assets is a defining responsibility of leaders of society's most powerful institutions. In the United States, we talk about the health of the economy by referencing the recent Dow Jones Industrial Average numbers. Our sense of the economy does not interrogate how people or the Earth are doing; instead, it relies on financial metrics like jobs numbers or unemployment data, posing as objective measurements of economic health.

We are assured there is no need for concern about the resulting inequality because the invisible hand of the market will distribute benefits according to merit, and all will eventually enjoy limitless material abundance—if they have earned it. It is shocking that a story so obviously flawed could be allowed to harm so many for so long without having sparked rebellion and corrective action. It is all too rare, however, that we educate our young to question the stories that define our lives and our communities in this way. Such challenges only arise if people venture outside of their communities and engage with those who view the world through different stories.

Successfully transforming our relationships with one another and the Earth requires a new economics grounded in an accurate story that is compelling. That story must be one that focuses our attention on securing the wellbeing of all people and our planet, treats money as a tool rather than a purpose, and reminds us that most of the real wealth of the living Earth is the product of *all* of life's labor. Once we get our story right, we have a chance to correct our priorities.

Defining Twenty-First-Century Priorities

A viable human future depends on redirecting our priorities relating to purpose, power, and procreation.

1. *Redirect Purpose from growing GDP to securing the wellbeing of people and Earth.* Living beings grow, but only within continuing cycles of birth and death. If our human body continues to grow past adolescence, it generally means we need to change our diet and get more exercise. In making perpetual GDP growth our defining human purpose, we have made money our defining value and created an economy at odds with our nature as living beings. We now have overwhelming evidence that growing GDP can be detrimental to the wellbeing of people and Earth.

    Imagine flying an airplane on a dark cloudy night with only an airspeed indicator and an instruction to keep the speed as high as possible. A steep dive is the best way to maximize speed. The result will be a crash. To safely fly an airplane with instruments alone requires multiple indicators, including airspeed, altitude, and direction.

    Kate Raworth, the widely acclaimed author of *Doughnut Economics*, suggests that managing a modern economy requires two indicator panels. One panel warns when essential human needs are not being met and the other warns when demands on Earth's critical regenerative systems are being exceeded. The goal is to meet the material needs of all people within the limits of Earth's regenerative capacity. Otherwise, we are nose-diving at an increasing speed.

2. *Redirect Power from money-seeking corporations to diverse life-serving bioregional communities.* Life exists only in communities of diverse organisms that self-organize to create and maintain the conditions

essential to their individual and collective wellbeing. For humans, this means organizing ourselves as substantially self-reliant bioregional communities that link urban and rural people, control the resources on which their wellbeing depends, and are accountable to their members.

Focused on money rather than life, our economies are dependent on the transfer of resources from living communities with vibrant local markets to monopolistic, profit-maximizing transnational corporations uncoupled from any responsibility or accountability to the communities in which they do business. The profit-maximizing, liability-limiting corporation is an illegitimate institution. These errant institutions must be broken up and restructured in ways consistent with democratic sharing of power and the responsibility of all human institutions to support community wellbeing.

3. *Redirect procreation from birthing more babies to assuring every child has a healthy, meaningful, and productive life.* The finite living Earth is the common creation and heritage of all living beings. The products of Earth's regenerative systems must be equitably shared with all human communities and other species throughout the Earth. No one holds a right to excess while others are denied their fundamental needs. The more we limit our human numbers, the more of Earth's products are available for us each to share. If we choose to grow our numbers, we have less to share, and must be more frugal.

Every child must be a wanted child who receives the loving care of family and community. No one should feel they must bear their own child to share in the joy of raising a child, or to assure their personal security in their final years.

The world has more than enough human children. What we lack is adequate attention to the care and development of all our children to their full potential. Likewise, we each have a right and responsibility to share in the joy and challenge of caring for the children on whom the human future depends. Our most essential individual contribution to posterity is not through the replication of our genes. It is through our contribution to the care, understanding, and wellbeing of the whole.

A New Economics for a New Civilization

The founders of economics envisioned a field of inquiry devoted to guiding the management of the human household. In line with that aspiration, an authentic economics for our twenty-first-century world will guide us to a future of meaningful and rewarding lives for all in sustainable balance with the living Earth. Its underlying story of human needs and aspirations will present a dramatic contrast to the story fabricated and promoted by twentieth-century economists. The table below contrasts the worlds that twentieth-century and twenty-first-century economists see through their contrasting conceptual lenses.

The Contrasting Lenses of Twentieth-Century and Twenty-first-Century Economics	
The Financial Story Lens of 20th-Century Economics	**The Living Earth Story Lens of 21st-Century Economics**
There are only individuals. Community is an illusion.	Life exists only in community. I am because you are.
There are no material limits to human consumption and economic growth.	Earth's regenerative systems are finite. We must live within their limits.
GDP is a defining measure of economic performance.	GDP measures growth in the exchange of money without regard to whether it is beneficial or harmful.
Financial return is the measure of an investment's benefit to society.	Productive investment and predatory speculation serve very different and conflicting ends.
Market price is an objective measure of the value of natural assets.	The value of the regenerative systems that maintain Earth's air, water, and soils is beyond price.
Economics is an independent science with little need to learn from other disciplines.	Economics for our troubled and changing world must be transdisciplinary and continuously learning.
Money is capital and the defining constraint in addressing society's needs. Society's wealth grows as its money grows.	Money is a number that has value only as a token of exchange for things others offer for sale. Central banks create it with a mere computer keystroke.
The owners of financial assets are society's wealth creators and should be rewarded accordingly.	Those who do productive work are society's real wealth creators and should be rewarded accordingly.
Reducing all values to money allows for quantification and makes economics an objective science.	An economics that values life only for its market value is an ideology, not a science.
Humans by nature are individualistic, competitive, aggressive, and self-serving.	Mentally healthy people are caring, peaceful, and derive pleasure from helping others.

| We all do best when we compete to maximize our personal advantage. | We all do best when we look out for and care for one another. |
| Limitless material consumption is the path to human happiness. | Material consumption beyond sufficiency is a distraction from meaning and true happiness. |

Viewing the world through a financial lens, the twentieth-century economist sees an economy defined by financial flows and seeks opportunities to grow GDP by increasing the size and efficiency of those flows. He or she favors empowering transnational corporations because of their ability to grow consumer demand, shareholder profits, and economic efficiency. These corporations organize long supply chains that connect specialized suppliers in distant countries in order to take advantage of low wages, lax environmental regulations, tax havens, and public subsidies. He or she sees these corporations as free market champions who are battling interference from governments and politicians seeking to impose taxes, tariffs, and regulations that limit the free market's ability to self-organize in the cause of efficiency and growth.

A viable human future depends on finding our way to a new civilization guided by the more complex and nuanced view of the world that is offered in the lens of twenty-first-century economics above. Through that lens we see the wondrous processes of life's miracle of self-organization as communities of the whole. We also see the ways in which financial flows facilitate and disrupt the natural processes on which all life ultimately depends.

Recognizing that GDP tells us nothing about whether growth in financial exchange is beneficial or harmful to people and planet, the twenty-first economist realizes it is absurd to make GDP growth a defining societal goal. He or she knows: life's wellbeing depends on resilient communities that control their resources to meet essential daily needs; and that the only legitimate purpose of any human institution—including those of business and government—is to serve the wellbeing of people and Earth. To meet this requirement, markets need rules created and enforced by democratically accountable governments.

⁓

A great many of us recoil from the thought of a return to business as usual after the COVID-19 pandemic. That reaction indicates that we may be on the verge of committing to truly transformational change. Our future depends on reconnecting to life and the whole of Earth's community with a deepened understanding of ourselves and our relationship to each other. To get our future right, we must get our economics right. To get our economics right, we must get our story right. The wellbeing of people and Earth, not GDP, must be our goal. Interdependence and mutual care—not individualism and competition—must define our relationships.

References

Elliot, Larry. "World's 26 Richest People Own as Much as Poorest 50%, Says Oxfam." *Guardian*, January 20, 2019. https://www.theguardian.com/business/2019/jan/21/world-26-richest-people-own-as-much-as-poorest-50-per-cent-oxfam-report.

Global Footprint Network. "Earth Overshoot Day 2019 is July 29th, the Earliest Ever." Accessed August 9, 2020. https://www.footprintnetwork.org/2019/06/26/press-release-june-2019-earth-overshoot-day/.